Secrets of

Making and

Breaking

Codes

HAMILTON NICKELS

Secrets of

Making and

Breaking

Codes

A Hands-on Guide to Both Simple
and Sophisticated Codes to Easily
Help You Become a Codemaster

Skyhorse Publishing

Skyhorse Publishing books may be purchased in bulk at special discounts for sales promotion, corporate gifts, fund-raising, or educational purposes. Special editions can also be created to specifications. For details, contact the Special Sales Department, Skyhorse Publishing, 307 West 36th Street, 11th Floor, New York, NY 10018 or info@skyhorsepublishing.com.

Skyhorse® and Skyhorse Publishing® are registered trademarks of Skyhorse Publishing, Inc.®, a Delaware corporation.

Visit our website at www.skyhorsepublishing.com.

10 9 8 7 6 5 4 3 2 1

Library of Congress Cataloging-in-Publication Data is available on file.

ISBN: 978-1-62914-206-7

Printed in China

To Clara J.

CONTENTS

CHAPTER 1

General Information

■

Secrets of Making and Breaking Codes is intended to be a practical, field-level manual designed to teach the comnat and survival professional the basic mechanics and methods of enciphering and deciphering communications.

I have attempted to limit information about the history and personalities in cryptography in order to present the maximum amount of practical material. Although codebreaking and cryptanalysis are touched on only lightly, the alert, interested student will be able to develop insight into the construction of code and cipher systems and, with experience, break some ciphers.

A wide range of ciphers and codes is covered, enabling the user to select a method or system that suits the requirements of the situation.

Sufficient materials are included so that the student can become proficient, then go on to teach others the methods.

All ciphers fall into the following categories:

1. **Concealment.** The simplest and most primitive, a concealment system attempts to hide (kryptos is a Greek word

meaning "hidden") the secret message, most often within another seemingly harmless message. But the secret message need not be written—it could be the arrangement of items in a store window or colors used in a painting. In fact, any device at all can be used in any way to convey information in the concealment message. In many cases the concealment message is passed to the targeted receiver in such a way that there is no indication that a message of any sort has been sent. In this manual, however, all systems presented can be reduced to a message form using alphabetical or numerical characters that can be written or transmitted by voice or Morse code.

2. **Transposition.** In transposition ciphers, the elements of the *plaintext* (see definitions at the end of this chapter) message, that is, the letters or symbols in the original text, are manipulated out of their normal order—they are jumbled and rearranged. Quite often there is a periodic pattern to the rearrangement, or the disarrangement might be controlled by a *key* that informs the intended receiver of the method to be used for deciphering the *ciphertext.*

3. **Substitution.** In a substitution system, the original message elements, letters, numbers, or other symbols are replaced with alternate symbols. If it is strictly a substitution cipher, the alternate or "stand in" symbols would be sent out in the *cryptogram* in the same order as the *plaintext* or original message elements. The schemes that have been developed for substitution-type systems are among the most complex and, if properly constructed, are virtually unbreakable. These systems are used in the highest levels of the military and by almost all national governments. These *cryptograms* are often sent openly to the targeted receiver without fear that the *enciphered* message will be *deciphered* by any adversary.

In reality, practically all systems are some mixture of the three fundamental systems and can be termed *product* ciphers. For example, a complete system may use a substitution of symbols, then transpose those new symbols, finally concealing them in a microphotograph—hoping to pass them to the intended receiver without alerting an adversary that any message has been sent.

CODE VERSUS CIPHER DEBATE

In the terminology section at the end of this chapter, you will notice that the word *code* has a special meaning to the professional cryptologist. He thinks of *code* and *cipher* as being two different things. In this manual, however, and in most general usage, the word *code* is used to mean a cipher or a list of words represented by numbers or other symbols or *codewords*. In most cases a *code*, as defined by the professionals, is so large that it is made up in book form.

This manual, in later chapters, has a workable *code* that has been used and tested in the field. You can tailor this *code* to use within your organization and maintain a certain degree of security, even if your cryptograms are intercepted and decrypted. A *code* can be enciphered. When this is done, it is said to be *superenciphered*.

SOME RULES

Nowhere in this manual will you find a statement that guarantees the security or safety of any given system. Over the years this promise has been made many times, yet even some of the most self-assured codemakers have been bested by even sharper codebreakers. To be safe, take the position that there is no such thing as the unbreakable system. If the professional codebreaker has sufficient messages, time, facilities, and motivation, the systems—that are practical for use in the field—can and will be broken. An exception is the one-time cipher. It is absolutely unbreakable. Other systems are classified as *virtually unbreakable*.

Rule #1: *Take for granted that the system you use can and will be broken.* Never be lulled to sleep by the promise of a new, simple system that is unbreakable.

Rule #2: *The system you use must be secure enough to delay* the codebreaker until the value of the message is worthless. Since most messages become worthless after a period of time, your most important decision as a cryptographer is to select and use a system that avoids or delays solution by an adversary until well after the value of the message has fallen to zero.

Rule #3: *The more you use a system, the greater the likelihood that the system will be broken.* Restated, change the system you are using as often as possible. Keep messages short.

Rule #4: *Destroy all cipher material, preferably by burning, then powdering the ash.*

Rule #5: *Ciphering should be done on hard-surfaced tables so that no impressions of pencil marks are left.*

Rule #6: *The plaintext and the enciphered message should not be written or recorded on the same paper, tape or disk.*

Rule #7: *Typewriter ribbons, computer discs, or tapes must be totally destroyed after each session* (or removed from machines and securely stored with cipher materials). Remember that "deleted" and "erased" material can be easily recovered from computer discs and tapes that are thought to be "clean" unless overwriting is done.

Rule #8: *Communications channels that are used for cryptograms must be used in a consistent fashion.* That is, null messages must be sent regularly so that no change in traffic can be noticed when these channels are needed for actual messages.

Rule #9: *All cipher materials must be accounted for and checked periodically.*

Rule #10: *Knowledge of the systems in use should be restricted to those professionals who need to know.*

TERMINOLOGY

Cipher. Also spelled **cypher.** Abbreviated by the symbols "C:" in this manual. See *code.* In the strictest sense, *cipher* refers to systems that manipulate one, two, or (rarely) three characters at a time. A letter pair is called a bigram; a triplet is a trigram. If a group of characters consists of more than two, it is a polygram. A *cipher* might look like this:

Cipher	Plaintext
d	A
z	B
5	C
j	D

C: Abbreviation for ciphertext or codetext, code, or cipher.
P: Abbreviation for plaintext or cleartext.
K: Abbreviation for key, keyword, keynumber.

Cipher Alphabet. Used in *substitution* systems, the cipher alphabet is a list of equivalents for the alphabet used by the language of the *plaintext*. Example:

plaintext
alphabet: ABCDEFGH I J KLMNOPQRST UVWXYZ
cipher alphabet: z o i u y t r e wq l k j h g f d s amn b v c xp

Ciphertext. Same as *codetext*.

Code. (See *cipher.*) In the strictest definition, *code* refers to a system which uses words as the smallest element. (In general usage, the word *code* and the word *cipher* are used interchangeably.) A *code* might look like this:

Codeword	Plaintext
angel	we are ready
fox	send additional
camera	advise position
blue	return at once

Codetext. The message after it has been *encoded*.

Concealment. Methods of hiding a message within another message.

Cryptanalysis, Cryptanalyze. The attempt to break down or solve a *cryptogram* by a person who does not have the *key* or proper method.

Cryptogram. The complete encoded message, with all its parts, in a form ready to be sent or already sent.

Cryptography. Any method used to change the form of a written message to make it difficult or impossible for anyone except the targeted receiver to read.

Decipher. Same as *decode* and *decrypt*.

Decode. The reverse of the *encoding* process that is performed by a receiver who knows the decoding method or holds the proper *key*.

Encipher. Same as *encode* and *encrypt*.

Encode. To process a message into a code or cipher. The message would then be called the *codetext* or *ciphertext*. Same as *encrypt* and *encipher*.

Key, Keyword, Keynumber, Keyphrase. (Abbreviated by the symbols "K:" in this manual.) An element of the message that controls the system's variables. The targeted receiver should have the key and know the method of application.

Monoalphabetic. An encoding system which uses only one *cipher alphabet*.

Null. A word, letter, or other symbol which has no meaning that is used only to fill a space in a message.

Plaintext. (Abbreviated by the symbols "P:" in this manual.) The message before it is *encoded*. The message after it is *decoded*. It is also called *cleartext* or *clear* or *open*.

Polyalphabetic. An encoding system which uses two or more *cipher alphabets*.

Steganography. Any method of transmitting a secret message that conceals the very existence of the message.

Substitution. A method of encoding using replacement of the elements in the *plaintext* by other letters, numbers, symbols, or by a combination of other elements.

Transposition. A method of encoding using disarrangement or shuffling of the order of the *plaintext* elements.

CHAPTER 2

Concealment

■

Hiding a message in a banana or in a shoe are examples of concealment systems, but in this manual, we are concerned with systems that conceal a message within another message. Consider the following:

I WILL DO IT. IF NOT TODAY, THEN TRUST ME. SIGNED, SMITH.

Can you read it? Copy out every third word.

I WILL DO IT. IF NOT TODAY, THEN TRUST ME.
      DO   NOT   TRUST
SIGNED, SMITH.
      SMITH
DO NOT TRUST SMITH

The skilled codebuster would look at the slightly awkward style of the message and bells would ring. He would look for a set of words that made more sense than the apparent message. In a matter of minutes he could break the code. Now look at this message:

EFDH GORA NQBO PETE YTDS RTOU ZESV ITVE SOWM XNIM CTLK HJEA

Can you decipher it? Copy out every third letter.

```
EFDH GORA NQBO PETE YTDS RTOU ZESV ITVE
   D    O    N    O   T    T    R    U   S    T
SOWM XNIM CTLK HJEA
  S    M    I    T   H
```

This cipher would take the professional a little longer, but he would have it broken in a very short time. Try this:

ADDE OAQO NPCR OOLL TMAT RLOC RATS TKCL
MNRA KETI SSTU ARTF THEE OSET ULCO JEOU TAKE
BLFZ IAHF SQUI TIFC ANLL TMZX AEXE DLGY ZZTI
FLOO VWKA TTIM IFTT HATH EEFC ANND FLHA

Read this by copying out a progressive cipher. Progress 1, 2, 3 . . . :

```
ADDE OAQO NPCR OOLL TMAT RLOC RATS TKCL
   D    O    N    O    T         T
MNRA KETI SSTU ARTF THEE OSET ULCO JEOU TAKE
   R         U         S                   T
BLFZ IAHF SQUI TIFC ANLL TMZX AEXE DLGY ZZTI
         S                   M                      I
FLOO VWKA TTIM IFTT HATH EEFC ANND FLHA
              T                              H
```

All of these systems use *nulls* (meaningless symbols) to fill in and separate the *plaintext*. In the last code group FLHA above, the F, L, and A, for example, are *nulls*.

Is the concealed message too easy to break? Try the

Concealment

following cryptogram sent on December 13, 1944, during World War II. A key was used.

   FANAB IASSO NADME ETSSH IPOWI LLNRU
NNLTO SOONK NIGIT NTTRE LONCE GCODE
LNOTE AISAN SFOUR

Remember that the strict definition of a concealment cipher does not allow shifting symbols or making substitutions. The message is there for you to see, but some method has been used to draw attention away from it. Without the *key*, even a professional cryptanalyst would struggle with this for some hours, perhaps longer.

```
FANAB IASSO NADME ETSSH IPOWI LLNRU
 A    B A    N D            O     N
NNLTO SOONK NIGIT NTTRE LONCE GCODE
  L O   O K   I N            G
LNOTE AISAN SFOUR
L     A S   S
```

At first glance, there doesn't seem to be any order in the arrangement of the nulls, but if you change the date into all numbers, December 13, 1944, becomes 12 13 1944 or 12131944, and you can see that the number of nulls between the plaintext corresponds to the value of each succeeding digit in the date.

```
F A N A B   I A S S O   N A D M E   E T S S H
1 A 2 2 B   1 A 3 3 3   N 1 D 9 9   9 9 9 9 9
I P O W I   L L N R U   N N L T O   S O O N K
9 9 O 4 4   4 4 N 4 4   4 4 L 1 0   2 2 0 1 K
N I G I T   N T T R E   L O N C E   G C O D E
3 3 3 I 1   N 9 9 9 9   9 9 9 9 9   G 4 4 4 4
L N O T E   A I S A N   S F O U R
L 4 4 4 4   A 1 S 2 2   S X X X X
```

The *key* was known both by the sender and the targeted receiver. The key did not have to be transmitted with the cryptogram, memorized, or written anywhere; it was the date of the message.

The very simplicity of a well-enciphered concealment message may confound the experienced codebuster for several hours, perhaps days, especially if he has only one message as a sample, but this is not a method to be used in a long-term situation. In general, simple concealment systems should be avoided in practical use. Transposition and substitution methods offer much more security and can be enciphered and deciphered in a comparable amount of time with about the same effort.

CHAPTER 3

Transposition Systems

■

Below is a very simple transposition cipher. See if you can decipher it.

C: NRUTE RAMOH ALKOL ECNAC

Everything is written backwards, then put into five-letter code groups. Decrypted it reads:

P: CANCE LOKLA HOMAR ETURN
P: CANCEL OKLAHOMA RETURN

Here is a form of *alternate-letter* transposition. The plain-text was:

P: RETURN TO BASE

It is enciphered by dropping every other letter to the second line.

```
R   T   R   T   B   S
  E   U   N   O   A   E
```

Then the second line is added to the end of the first line:

C: RTRTBS EUNOAE

and put in four-letter code groups:

C: RTRT BSEU NOAE

Encoding, decoding, and transmitting a message is easier if it is in code groups. This treatment makes it somewhat more difficult for the adversary to break. Four- and five-letter code groups are most often used.

If you need to send numbers or other symbols, it may be best to send them written out as in this example:

P: CHARLIE WILL ATTACK 2100

Change to:

P: CHARLIE WILL ATTACK TWO ONE ZERO ZERO

The message was encrypted by first making a rectangle of 5×7 letters out of the plaintext or *clear*, as it is sometimes called. In this case, it was agreed beforehand that the number of letters sent in a code group would be equal to the number of letters in each *row*. Then, the clear would be arranged by writing from left to right in the first row, then down to the second row as in conventional writing as seen in the 5×7 box on the following page:

```
·  1 2 3 4 5
1 C H A R L
2 I E W I L
3 L A T T A
4 C K T W O
5 O N E Z E
6 R O Z E R
7 O
```

It had also been agreed that the letters would be encrypted (transposed) by *taking out* or *taking off* the letters by going down the first column on the left, then *taking off* down column 2, then 3, 4, 5. Nulls AEIO would be added at the end of the message to make the rectangle a full 5×7. The message sent was:

C: CILCO ROHEA KNOAA WTTEZ ERITW ZEILL AOERO

Decrypted by reversing the encrypting procedure:

P: CHARL IEWIL LATTA CKTWO ONEZE ROZER OAEIO
or:
P: CHARLIE WILL ATTACK 2100

Here we will encipher the same message using a more elaborate transposition system. This scheme was used by several nations as their military and diplomatic systems some years ago. A *keyword* is used and nulls are worked into the ciphertext. First, the clear is arranged as in the earlier example.

```
CHARL
IEWIL
LATTA
CKTWO
ONEZE
ROZER
O
```

Then, nulls are inserted in a prearranged fashion.

CHARL
AABBC
IEWIL
DDEEF
LATTA
GGXXA
CKTWO
MNOPQ
ONEZE
RSTUV
ROZER
WXYZA
OAEIO

The nulls are carefully chosen to mislead the adversary crypt-analyst to make him think that it is perhaps a substitution-type code. To make the keyword easy to remember, we are using FRANK. A word that has five letters with no repetitions is chosen. The targeted receiver must know the keyword that will be used in messages sent to him.

FRANK is converted to a numerical form by assigning numbers to the letters according to their order of appearance in the alphabet. Then the numbers are used to determine the order that will be followed in taking off the columns as they are enciphered. The numerical value for FRANK would be 25143. This is used as the heading for the rectangle of letters containing the clear and the nulls.

K: F R A N K
K: 2 5 1 4 3
   C H A R L
   A A B B C
   I E W I L
   D D E E F
   L A T T A
   G G X X A
   C K T W O
   M N O P Q
   O N E Z E
   R S T U V
   R O Z E R
   W X Y Z A
   O A E I O

The column 1 under A in FRANK will be taken out first, then column 2 under the F, then K, then N, then R.

C: ABWETXTOETZYE CAIDLGCMORRWO LCLFAAOQEVRAO RBIETXWPZUEZI HAEDAGKNNSOXA

Then arrange into five-letter code groups.

C: ABWET XTOET ZYECA IDLGC MORRW OLCLF AAOQE VRAOR BIETX WPZUE ZIHAE DAGKN NSOXA

We will go through the decoding to check for mistakes. This should be done with all cryptograms, preferably by another person who has not participated in the encoding. Here is the cryptogram:

C: ABWET XTOET ZYECA IDLGC MORRW OLCLF AAOQE VRAOR BIETX WPZUE ZIHAE DAGKN NSOXA

We know from the keyword and by prearrangement that there are five columns. Count the number of letters in the message: 65. Divide by five and you have the number of letters in each column: 13.

Write out FRANK and assign numbers relative to each letter ranking it according to its place in the alphabet.

K: F R A N K
K: 2 5 1 4 3

Take the first 13 letters of the codetext and put them under the "1" column.

K: F R A N K
K: 2 5 1 4 3
        A
        B
        W
        E
        T
        X
        T
        O
        E
        T
        Z
        Y
        E

Now take the next 13 letters and place them under column 2.

```
K: F R A N K
K: 2 5 1 4 3
    C   A
    A   B
    I   W
    D   E
    L   T
    G   X
    C   T
    M   O
    O   E
    R   T
    R   Z
    W   Y
    O   E
```

Do the same for columns 3, 4, and 5.

```
K: F R A N K
K: 2 5 1 4 3
    C H A R L
    A A B B C
    I E W I L
    D D E E F
    L A T T A
    G G X X A
    C K T W O
    M N O P Q
    O N E Z E
    R S T U V
    R O Z E R
    W X Y Z A
    O A E I O
```

Now remove the nulls filling every other row, as was prearranged.

```
K: F R A N K
K: 2 5 1 4 3
   C H A R L
   I EW I L
   L A T T A
   C K TWO
   O N E Z E
   R O Z E R
   O A E I O
```

Read off the plaintext from left to right, then down, as in the encoding procedure.

P: CHARL IEWIL LATTA CKTWO ONEZE ROZER OAEIO

Remove the nulls at the end.

P: CHARLIE WILL ATTACK TWO ONE ZERO ZERO
or:
P: CHARLIE WILL ATTACK 2100

Hundreds of variations of columnar transposition schemes have been used successfully with considerable safety. Repeated use of the same form, or using the same keyword, however, would make the system vulnerable to the codebreaker.

Here is a relative of the columnar type that is extremely easy to use. The sender and the targeted receiver must know the keyword. A keyword is chosen that is relatively long yet has no repetitive letters. For this cryptogram MICKEY was used. This message was sent on June 4, 1944:

C: INEHU TIASS ITZOE

This was the message in the clear:

P: IT IS HOT IN SUEZ

As is often true, in this case, the clear was itself a part of a prearranged code. When a code or previous encipherment is enciphered into a new cryptogram, the message is said to be *superenciphered*.

The receipt of this prearranged code meant to the Resistance fighters in France that the Allied forces were about to invade Europe and that they should take immediate action to destroy all possible railway systems and equipment in Normandy.

This is how it was encoded. The plaintext had thirteen letters. Fifteen were needed for five-letter code groups. Two nulls, JX, were added to make the cryptogram appear to be a substitution type. The keyword MICKEY was repeated so that there were fifteen characters in it also. Each letter in the keyword was assigned a numerical value according to its rank in the alphabet, then its rank in the keyword itself. The numerical values were continued and extended as the keyword repeated.

```
K: M  I C K E Y  M  I C K  E  Y  M  I C
K: 11 6 1 9 4 14 12 7 2 10 5 15 13 8 3
K: I  T I S H O  T  I N S  U  E  Z  J X
```

Take out the letters from the plaintext and encode them according to the order of the numbers under the keyword.

```
C: I N X H U T I J S S  I  T  Z  O  E
   1 2 3 4 5 6 7 8 9 10 11 12 13 14 15
```

Break into five-letter code groups.

C: INXHU TIJSS ITZOE

This and hundreds of other messages that were sent—many of them made up entirely of nulls—apparently added to the rampant confusion that existed in German intelligence at this critical time, just before the Allied invasion in Normandy.

It is now well known that a number of high-ranking officers in the German command did not believe it possible that the Allies could land a significant force in Europe in 1944, and consequently they ignored many of the Allied forces' communications.

In fact, the Germans not only greatly underestimated the Allied forces' military capability but their intelligence people as well. The British had methodically obtained copies of the German naval codebooks, primarily by sending divers down at the sites of ship and submarine sinkings. The German ciphers were clever and many of them were never broken, but their naval codebook system was almost completely known to the Allies throughout the war. The British and the Americans, working together, finally broke important cipher systems that were being used primarily by the German submarine fleet. When this happened, it helped turn the tide against the Germans in the North Atlantic; the Americans sent ever-increasing streams of war materials to the British and Russians, and it was the beginning of the end for Germany.

It should be mentioned here, also, that the Americans had complete knowledge throughout the war of the Japanese code-machine cipher system called PURPLE. The Japanese were informed by numerous sources that their system had been violated, even by an article that ran in the *Chicago Tribune* and was later carried by many U.S. and foreign newspapers, but they stubbornly refused to believe it. They continued to use PURPLE until the end of the war. According to reliable sources, they never tested the system by sending false information that would cause the Americans to show their hand. But the really puzzling thing about the PURPLE affair is this: with complete knowledge of Japan's every move lying on the desk of the president of the United States each morning, why did it take so long to bring the war in the Pacific to a close? We will probably never know since most of the records of intelligence activity have been destroyed or will never be released to the public.

The PURPLE machine was actually a Japanese copy of the Enigma rotor machine that had been available in Europe in generic form for under a thousand dollars (U.S.) for a number of years. Tied to it were two American-made electric typewriters that allowed input and output to the rotors that mixed or substituted alphabets for each letter input. The general system was well known to the intelligence community of all of the major nations of the world. The primary advantage of these machines was that for a given keyword, literally thousands of substitution alphabets could be produced by the rotor system. Another advantage was that they allowed rapid message handling with few errors.

Today, we all have potential "code machines" around us in the form of computers and even in the ever-present $4.99 calculators.

Below is a difficult system to decipher that uses a calculator to obtain a complicated key quickly. In a later chapter we will construct a cipher with a calculator that is virtually unbreakable.

We will use, as an example, a message similar to one sent by Arthur Zimmermann, the German foreign minister, to Mexico in 1917. German cryptographers used a cipher known to them as 0075. The message sent was:

C: CTLTZ EMRTH IERSI TNAII WETXC AAMOR
OXCEA ATWOA AONIZ NEETN MXASA LDINF
ESZRC ATEIO GZFXA LAEIR AOMBI OWEWW

Unfortunately for Zimmermann and the Mexicans, British intelligence cracked 0075 and wired U.S. President Woodrow Wilson the following information (paraphrased here):

CONTENTS OF ZIMMERMANN CABLE FOR YOUR INSPECTION:
P: "CONFIRM THAT MEXICO WILL BE AWARDED TITLE TO ARIZONA TEXAS NEW MEXICO IF MEXICO ENTER WAR AGAINST USA AZ AZ AZ"

Here is the method used to encipher the message. The key, 0075, is used in a simple equation to obtain the *control key*.

$$K: \frac{19999 + KEY}{97} = CONTROL\ KEY$$

$$K: \frac{19999 + 0075}{97} = 206.9484536$$

All digits that appear on the screen of the calculator are used and the decimal point is ignored.

K: 2069484536

This series of digits is *ranked* according to the value of each digit and its place in the series. Zero is counted as the value 10.

K: 2 0 6 9 4 8 4 5 3 6   CONTROL KEY
K: 1 10 6 9 3 8 4 5 2 7   RANKED CONTROL KEY

The plaintext is now written below the ranked control key, ten letters in each line, but note that they are *written in* (as opposed to *taken out*) according to the ranked control key.

K: 1 10 6 9 3 8 4 5 2 7
  C A R H N T F I O M   (C O N F I R M T H A)
  T L C I E W X I M O
  L D A E E D A W B R
  T I T R T A L E I O
  Z N E S N A A T O X
  E F I I M O E X W C
  M E O T X N I C E E
  R S G N A I R A W A
  T Z Z A S Z A A U A
  1 2 3 4 5 6 7 8 9 10

Next, notice that we have written below the columns a new rank. In this case, to keep it simple, we are using the normal order of numbers from 1 to 10. You could, for variation, reverse the order of the ranked control key or use a reversal of the normal order of numbers. The variations are almost limitless. In this case, however, we proceed to take out the letters in our new column 1, writing them from left to right.

C T L T Z EMR T

The columns ranked 2, 3, 4, 5, 6, 7, 8, 9, 10 follow in order.

C: CTLTZEMRT ALDINFESZ
   RCATEIOGZ HIERSITNA
   NEETNMXAS TWDAAONIZ
   FXALAEIRA IIWETXCAA
   OMBIOWEWW MOROXCEAA

Now the ciphertext is divided into five-letter words.

C: CTLTZ EMRTH IERSI TNAII WETXC AAMOR
   OXCEA ATWOA AONIZ NEETN MXASA LDINF
   ESZRC ATEIO GZFXA LAEIR AOMBI OWEWW

Note that we have actually done a double transposition in the above example.

Transposition ciphers are simple, quick, and valuable for use in the field. But beware—the primary drawback is that the message elements are completely exposed. In the above cryptogram, the known fact that the targeted receiver was Mexico, coupled with the repetition of the numerous X characters, would tip off an experienced cryptanalyst that the plaintext word MEXICO must certainly be present. Ordinarily, a single *known word* or *probable word* is all that is needed to crack this type of cipher. Of course, it would take time, and on the field level most messages have a short time value.

POCKETRAN SYSTEM

A tougher, virtually unbreakable transposition system, developed by the author in 1974, uses a computer to select a pseudo *one-time key*. The concept, as will be seen and expanded upon in later chapters, constitutes one of the few reliable methods to produce a virtually unbreakable ciphertext in the field. Here is a message that was sent in 1975 using a system similar to the program below:

C: PIIZJ KTROF AXALM EVDYU GWNBW CQSDH

The plaintext was:

P: WIGWA MISRE ADYBC DFHJK LNOPQ TUVXZ
or:
P: WIGWAM IS READY

The key used here was 110275. This was the date the message was sent and received.

What puts this system in the virtually unbreakable category? First, the disarrangement of the letters in the plaintext is controlled by a series of carefully calculated numbers generated by the computer. Second, over half the letters in the plaintext are nulls but are chosen to constitute the entire alphabet. By including these particular nulls, more than 200,000,000,000,000,000,000,000,000,000,000 (200 nonillion) different combinations would have to be inspected by a computer running programs to break the cipher—more words by far than there are in all the languages that have ever been spoken on planet Earth. To get a real idea of the magnitude of that number, assume that a computer could run a codebuster program, produce a possible word, check it against a dictionary, and print it—all in one second. To run all combinations would require over 8,000,000,000,000,000,000,000,000 (8 septillion) years.

Further, the system can accept virtually any positive or negative whole number or fractional number and use it as a key. So if we calculated the number of different possible keys, we would have a number larger than those above—too large to think about.

Below is a listing of the program in the BASIC computer language using the dialect of the Tandy PC-7 pocket computer manufactured by Casio. Because of the pocket-sized two-kilobyte memory space of the PC-7, this version will only encipher or decipher a message limited to thirty characters. Another minor problem of the PC-7 is that it takes three or more minutes to run through the ten to twenty thousand calculations required to generate a unique encipherment sequence for each key. Really not so bad when you consider that it would take a human more than seven days to do it on a calculator—provided the human made no mistakes.

POCKETRAN PROGRAM

```
20 CLEAR : DEFM 66 : GOTO 700
40 INPUT "NEW KEY " , K
50 PRINT "WAIT"; : REM COPYRIGHT 1986, 1989
   H. NICKELS
60 Z = K * 3.1416 : N = 7 : M = 2 : REM
   POCKETRAN V6.2C
80 FOR A = 31 TO 60
100 IF N = 300 THEN M = M + .5 : N = 3
110 N = N + 1
120 Z = INT(FRAC(((997 * Z + N↑M)/199)) * 100)
140 IF Z > 30 THEN GOTO 100
160 IF Z < 1 THEN GOTO 100
180 B(A) = Z
185 T = A - 1
200 FOR R = 30 TO T
220 IF Z = B(R) THEN GOTO 100
240 NEXT R
260 NEXT A
265 PRINT : GOTO 710
270 INPUT "MODE E OR D " , X$
```

```
272 IF X$ > "E" THEN GOTO 270
274 IF X$ < "D" THEN GOTO 270
276 IF X$ = "D" THEN GOTO 500
280 INPUT "PLNTXT" , $
300 FOR S = 31 TO 60
320 PRINT MID$(B(S),1)
340 NEXT S
342 PRINT "MESSAGE ENCIPHERED",
360 GOTO 720
500 INPUT "CPHRTXT ",$
520 FOR W = 31 TO 60
540 B$(60 + B(W)) = MID$(W – 30,1)
560 NEXT W
580 FOR V = 61 TO 90
600 PRINT B$(V)
620 NEXT V
630 PRINT "MESSAGE DECIPHERED", : GOTO 740
700 INPUT "NEW KEY Y/N",C$ : IF C$ = "Y" THEN
    GOTO 40
710 INPUT "CIPHER Y/N",C$ : IF C$ = "Y" THEN
    GOTO 270
720 INPUT "CK ENCPHR Y/N",C$ : IF C$ = "Y"
    THEN GOTO 300
740 INPUT "CK DECPHR Y/N",C$ : IF C$ = "Y"
    THEN GOTO 520
760 GOTO 700
```

The program will also run on the Tandy PC-4 and PC-6 models, various Casio and Sharp pocket computers, or, in fact, any computer that has BASIC. Minor adjustments must be made in some cases for the variations in syntax.

Below is a version written in BASIC that can be used on IBM compatibles. In this case, to illustrate another method, a keyword is used that is examined by the computer in two different ways. First, the numerical value of each letter is determined, summed, then manipulated by an internal key. The first number obtained from this is used as a seed number to start the number generator at the same point each time the same keyword is used. A second number, derived in a different way from the keyword, is used to step a certain point into the sequence of numbers that is generated. Then, only numbers from that point on will be used to determine the final disarrangement scheme. Sounds complicated, but the computer does all the work for you in a few minutes.

To keep this program short and relatively understandable, the screen display has been left in rudimentary form and the message length has been kept to thirty characters. In later chapters, we will develop a number of sophisticated versions that have text editors (word processors) and are able to handle messages of several thousand characters.

POCKETRAN FOR IBM/PC COMPATIBLES

```
10 CLEAR
20 FLAGA = 2
30 INPUT "ENTER MASTER KEY NUMBER (2 TO
   1000) DEFAULT IS 1 ",M : CLS
40 DIM A (30) : DIM B(30) : DIM C(30) : DIM D(30) :
   GOTO 120
50 ' COPYRIGHT 1989 H. NICKELS VERSION 6.30PC
60 CLS : LINE INPUT "ENTER PLAINTEXT (30
   CHARACTERS) ";P$
70 FOR X = 1 TO 30
80 A(X) = ASC(MID$(P$,X,1))
90 NEXT X
100 GOTO 400
110 '
120 INPUT "KEYWORD ";K$ : CLS
130 FOR AKEY = 1 TO LEN(K$)
140 D(AKEY) = ASC(MID$(K$,AKEY,1)) : ' PRINT
    D(AKEY),
150 TOTKEY = TOTKEY + D(AKEY) : Z = TOTKEY :
    ' PRINT Z,
160 NEXT AKEY
```

```
170 '
180 PRINT
190 FOR P = 1 TO ASC(K$)
200 N = N + 1: F = N*N : Y = (997*Z + F)/199 : Z = (Y –
    INT(Y)): Z = INT(Z*100)
210 ' PRINT Z,
220 NEXT P
230 '
240 FOR Q = 1 TO 30
250 M = M + 1: F = M^FLAGA : Y = (997*Z + F)/199 :
    Z = (Y – INT (Y)): Z = INT(Z*100)
260 IF M = 500 THEN FLAGA = FLAGA + .5 : M = 1
270 IF Z > 30 OR Z < 1 THEN GOTO 250
280 B(Q) = Z
290 FOR R = 1 TO Q – 1
300 IF Z = B(R) THEN GOTO 250
310 NEXT R
320 ' PRINT Z,
330 NEXT Q
340 M = 0
350 '
360 PRINT : INPUT "(E)ncipher or (D)ecipher ";C$
370 CLS : IF (C$ <> "E") AND (C$ <> "D") THEN 360
380 IF C$ = "E" THEN 60 ELSE 490
390 '
400 PRINT : PRINT "CIPHERTEXT"
410 FOR S = 1 TO 30
420 C(B(S)) = A(S)
430 NEXT S
440 '
450 FOR T = 1 TO 30
460 PRINT CHR$(C(T)), : LPRINT CHR$(C(T)),
470 NEXT T
480 END
490 '
500 PRINT : LINE INPUT "ENTER CIPHERTEXT (30
    CHARACTERS) ";Q$
```

```
510 FOR U = 1 TO 30
520 C(U) = ASC(MID$(Q$,U,1))
530 NEXT U
540 '
550 FOR V = 1 TO 30
560 A(V) = C(B(V))
570 NEXT V
580 '
590 PRINT : PRINT "PLAINTEXT"
600 FOR W = 1 TO 30
610 PRINT CHR$(A(W)), : LPRINT CHR$(A(W)),
620 NEXT W
```

(See Chapter 8 for versions of these programs that are a bit more secure and have features allowing general field use.)

CHAPTER 4

Basic Substitution Systems

All language, writing, and other symbols that man uses to communicate are codes. Words, sentences, and symbols are substitutes for ideas, actions, or objects. The human brain, in most cases, efficiently encodes and decodes the messages sent and received by our expressions and senses.

Of the various methods of communication used by man, writing seems to be the first of the "secret" codes. Historically, early writing was most often found in the possession of the ruling classes: priests, politicians, military forces, and merchants. In some cases it was a crime to teach anyone other than a member of one of these groups to write and read. Actually, little has changed. Today these same forces possess the best systems of secrecy.

It seems that as soon as man learned to communicate, secret systems began to develop, especially involving the practice of writing. Even in the earliest writing, cryptologists have found evidence of variant symbols being substituted for the common symbols.

Closer to the present day, for generations the European

ruling classes communicated in "foreign" or "dead" languages. As an example, lacking a practical code system at any moment, the British officer, when communicating to another of his kind, could always invoke the use of Latin learned in his required schooling.

Even today, physicians, attorneys, and other professionals still cling to and enlarge the system of the substitution of Latin, Greek, and "technical" words and phrases in order to keep their clients and other outsiders from knowing exactly what they are about. In earlier times, most of the disguised messages were of a political or military nature, but today, every profession, trade, or occupation—for example, the field of cryptology—has developed its own jargon, which is nothing more than the substitution of secret words for the more commonly understood ones. These words serve to keep the general public out and the insiders in. Probably the first thing—the most important thing—that anyone must learn to enter a new profession is simply the jargon or secret code of that profession or occupation.

In some cases a code is not particularly meant to deceive anyone but is a useful adaptation of the common language. It is a substitution of necessity. For example, the dits and dahs of the familiar Morse code—while they serve to keep the general public from eavesdropping on the broadcasts and transmissions of the users—are well known among a large group of people who can easily decode them. Consider speed-writing and shorthand. This is a system that allows the user to encipher the spoken word in a quicker, more efficient code than longhand. And people who are not able to hear and/or speak use a code made of hand signs that enables them to communicate.

The systems above are all familiar substitution systems, but they are not, strictly speaking, ciphers. Ciphers are ordinarily thought of as single, double, or triple groups of letters of the alphabet replaced with alternate symbols. These symbols may be anything. In all our cases, though, we will use only members of the English alphabet, since messages sent and received in a field situation must be carefully organized and standardized in order to avoid errors. In most cases, the use of symbols other than the alphabet does not enhance the security of the cipher system. A system such as this:

C: 53!!/305))6*;4826)4!.4!);806*;48/8 + 60))85;1!(;:!*8/83(8 8)5*/;46(;88*96*?;8)*!(;485);5*/2:*!(;4956*2(5* − 4)8 + 8*; 4069285);)6/8)4!!;1(!9;48081;8:8!1;48/85;4)485/528806*8 1(!9;48;(88;4(!?34;48)4!;161;:188;!?;

may at first seem to be extremely confusing, but to the cryptanalyst, the symbols used are irrelevant. If anything, it would benefit him because he has been given a hint by the use of these symbols that he is dealing with an amateur. Here is one way the cryptanalyst might approach the problem.

Make a table of the symbols with the number of times each appears:

Symbols	Number of Times Appearing
8	33
;	26
4	19
!	16
)	15
*	13
5	12
6	11
1	9
0	6
9	5
:	4
3	4
?	3
+	2
−	1

The cryptanalyst would then consult a frequency chart (see Appendix). E is the most frequently used letter—about 12 percent of the time. Below is a ranking of frequency of use of all letters used in normal text:

E, T, A, O, I, N, S, R, H, L, D, C, U, P, M, F, W, G, Y, B, V, K, Q, X, J, Z

Since E heads the list, substitute it for the symbol that is used most often in the cryptogram.

53!!/305))6*;4E26)4!.4!);E06*;4E/E + 60))E5;1!(;:!*E/E3(EE)5* /;46(;EE*96*?;E)*!(;4E5);5*/2:*!(;4956*2(5* – 4)E + E*;40692E5) ;)6/E)4!!;1(!9;4E0E1;E:E!1;4E/E5;4)4E5/52EE06*E1(!9;4E;(EE; 4(!?34;4E)4!;161;:1EE;!?;

T is the second-ranked letter in normal text. Try it for the second most-used symbol in the cryptogram.

53!!/305))6*T4E26)4!.4!)TE06*T4E/E + 60))E5T1!(T:!*E/E3(EE)5* /T46(TEE*96*?TE)*!(T4E5)T5*/2:*!(T4956*2(5* – 4)E + E*T40692E5) T)6/E)4!!T1(!9T4E0E1TE:E!1T4E/E5T4)4E5/52EE06*E1(!9T4ET(EET 4(!?34T4E)4!T161T:1EET!?T

Doesn't look like much help, but the cryptanalyst knows from the table of trigrams that the most common three-letter word is *the*. He notes that the combination T4E appears often throughout the cryptogram. The is the probable word he is looking for first. Substitute H for each 4 in the cryptogram.

53!!/305))6*THE26)H!.H!)TE06*THE/E + 60))E5T1!(T :!*E/E3(EE)5*/TH6(TEE*96*?TE)*!(THE5)T5*/2:*! (TH956*2(5* – H)E + E*TH0692E5)T)6/E)H!!T1(!9THE 0E1TE:E!1THE/E5TH)HE5/52EE06*E1(!9THET(EETH (!?3HTHE)H!T161T:1EET!?T

Now the cryptanalyst must run substitutions for the next most frequent symbol in the cryptogram, the !. The frequency table lists A as the next letter in line.

53AA/305))6*THE26)HA.HA)TE06*THE/E + 60))E5T1A(T
:A*E/E3(EE)5*/TH6(TEE*96*?TE)*A(THE5)T5*/2:*A
(TH956*2(5* – H)E + E*TH0692E5)T)6/E)HAAT1(A9THE
0E1TE:EA1THE/E5TH)HE5/52EE06*E1(A9THET(EETH
(A?3HTHE)HAT161T:1EETA?T

The occurrence of the double AA several times in the trial decipherment discourages the cryptanalyst. AA is seldom used in English. Further, no new recognizable words pop up except HAT. He decides to try the next highest frequency letter, O.

53OO/305))6*THE26)HO.HO)TE06*THE/E + 60))E5T1O(T
:O*E/E3(EE)5*/TH6(TEE*96*?TE)*O(THE5)T5*/2:*O
(TH956*2(5* – H)E + E*TH0692E5)T)6/E)HOOT1(O9THE
0E1TE:EO1THE/E5TH)HE5/52EE06*E1(O9THET(EETH
(O?3HTHE)HOT161T:1EETO?T

Now he sees OO, a very common double. He rejects II, NN, but SS is high in the single-letter list and in the doubles list.

53SS/305))6*THE26)HS.HS)TE06*THE/E + 60))E5T1S(T
:S*E/E3(EE)5*/TH6(TEE*96*?TE)*S(THE5)T5*/2:*S
(TH956*2(5* – H)E + E*TH0692E5)T)6/E)HSST1(S9THE
0E1TE:ES1THE/E5TH)HE5/52EE06*E1(S9THET(EETH
(S?3HTHE)HST161T:1EETS?T

This clearly looks wrong. No new words. No help. Decides to try the double SS on the) symbol since several occurrences of the double)) appear. The) is high in the count and S is high in the frequency list.

53OO/305SS6* THE 26SHO. HOST E06* THE
/E + 60SSE5T1O(T:O *E/E3(EES5*
/TH6(TEE*96*?TES*O(THE
5ST5*/2:*O(TH956*2(5* – HSE + E*TH0692E5STS6/E
SHOOT 1(O9 THE 0E1TE:EO1 THE /E5TH SHE
5/52EE06*E1(O9 THE T(EETH(O?3H THE SHOT
161T:1EE TO ?T

Progress. More new words. Separate the words from the rest
of the cryptogram. A is the third most often used letter in
English. Substitute it for the next unknowns. Try * first.

53OO/305SS6A THE 26SHO. HOST E06A THE
/E + 60SSE5T1O(T:O AE/E3(EES5A
/TH6(TEEA96A?TESAO(THE
5ST5A/2:AO(TH956A2(5A – HSE + EATH0692E5STS6/E
SHOOT 1(O9 THE 0E1TE:EO1 THE /E5TH SHE
5/52EE06AE1(O9 THE T(EETH(O?3H THE SHOT
161T:1EE TO ?T

No new words. Causes several occurrences of vowels con-
tacting vowels. Not encouraging. Try the next combination;
substitute A for the symbol 5.

A3OO/30 ASS 6* THE 26SHO. HOST E06* THE
/E + 60S SEAT 1O(T:O *E/E3(EESA*
/TH6(TEE*96*?TES*O(THE
AS TA*/2:*O(TH9A6*2(A* – HSE + E*TH0692 EAST S6/E
SHOOT 1(O9 THE 0E1TE:EO1 THE /EAT H SHE
A/A2EE06*E1(O9 THE T(EETH(O?3H THE SHOT
161T:1EE TO ?T

Looks better. Several new words. Find a place for I. Try I
for *.

Basic Substitution Systems

A3OO/30 ASS 6I THE 26SHO. HOST E06I THE
/E+60S SEAT 1O(T:OIE/E3(EESAI
/TH6(TEEI96I?TESIO(THE
AS TAI/2:IO(TH9A6I2(AI – HSE + EITH0692 EAST S6/E
SHOOT 1(O9 THE 0E1TE:EO1 THE / EAT H SHE
A/A2EE06IE1(O9 THE T(EETH(O?3H THE SHOT
161T:1EE TO ?T

No help. No new words. Further, I seems to be terminal—
not likely. N, the next letter in line, however, is often terminal.
Try N for *.

A3OO/30 ASS 6N THE 26SHO. HOST E06N THE
/E+60S SEAT 1O(T: ONE /E3(EES AN
/TH6(TEEN 96N?TES NO (THE
AS TAN/2: NO (TH9A6N2(AN – HSE + ENTH0692 EAST
S6/E SHOOT 1(O9 THE 0E1TE:EO1 THE / EAT H SHE
A/A2EE06NE1(O9 THE T(EETH(O?3H THE SHOT
161T:1EE TO ?T

Several new words. Notice a possible two-letter word in the
first line, 6N. It appears twice. Could it be IN? I is an excellent
candidate. It is high on the list but hasn't worked out yet
in the cryptogram. Try it.

A3OO/30 ASS IN THE 2 IS HO. HOST E0 IN THE
/E+I0S SEAT 1O(T: ONE /E3(EES AN /THI(TEEN
9 IN ?TES NO (THE AS TAN /2: NO (TH9A IN
2(AN – HSE + ENTH0I92 EAST SI/E SHOOT 1(O9 THE
0E1TE:EO1 THE / EAT H SHE A/A2EE0INE1(O9
THE T(EETH(O?3H THE SHOT 1I1T:1EE TO ?T

New words. Find a place for R, the next letter on the
frequency list. Replace (with it.

A3OO/30 ASS IN THE 2 IS HO. HOST E0 IN THE
/E + I0S SEAT 1ORT: ONE /E3REES AN / THIRTEEN
9 IN ?TES NORTH EAST AN /2: NORTH 9A IN
2R AN – HSE + ENTH0I92 EAST SI/E SHOOT 1RO9 THE
0E1TE:EO1 THE / EAT H SHE A/A2EE0INE1RO9
THE TREE THRO?3H THE SHOT 1I1T:1EE TO ?T

More new words. The second most common trigraph is
AND. Looking at the second line, there are two occurrences
of *AN/*. Could this be *AND*? Try substituting *D* for /.

A3OOD30 ASS IN THE 2 IS HO. HOST E0 IN THE
DE + I0S SEAT 1ORT: ONE DE3REES AND THIRTEEN
9 IN ?TES NORTH EAST AND 2: NORTH 9A IN
2R AN – HSE + ENTH0I92 EAST SIDE SHOOT 1RO9
THE 0E1TE:EO1 THE DEATHS HEAD A 2EE0INE1RO9
THE TREE THRO?3H THE SHOT 1I1T:1EE TO ?T

Where does *L* fit in? It is the next letter on the frequency
list. 0 is high on the list of used letters; try *L* for 0.

A3OOD3L ASS IN THE 2 IS HO. HOSTEL IN THE
DE + ILS SEAT 1ORT: ONE DE3REES AND THIRTEEN
9 IN ?TES NORTH EAST AND 2: NORTH 9A IN
2R AN – HSE + ENTHLI92 EAST SIDE SHOOT 1RO9
THE LE1TE:EO1 THE DEATHS HEAD A 2EE LINE 1RO9
THE TREE THRO?3H THE SHOT 1I1T:1EE TO ?T

Looks all right. More new words. *C* is the next letter in the
frequency series. Substitute it for 9, 2, and :. Do them all
at the same time.

A3OOD3L ASS IN THE C IS HO. HOSTEL IN THE
DE + ILS SEAT 1ORTC ONE DE3REES AND THIRTEEN
C IN ?TES NORTH EAST AND CC NORTH CA IN
CR AN – HSE + ENTHLICC EAST SIDE SHOOT 1ROC
THE LE1TECEO1 THE DEATHS HEAD A CEE LINE
1ROC THE TREE THRO?3H THE SHOT 1I1TC1EE TO ?T

The shotgun approach with the C did not work. Try the next
letter, U, using the same method.

A3OOD3L ASS IN THE U IS HO. HOSTEL IN THE
DE + ILS SEAT 1ORTU ONE DE3REES AND THIRTEEN
U IN ?TES NORTH EAST AND UU NORTH UA IN
UR AN – HSE + ENTHLIUU EAST SIDE SHOOT 1ROU
THE LE1TEUEO1 THE DEATHS HEAD A UEE LINE
1ROU THE TREE THRO?3H THE SHOT 1I1TU1EE TO ?T

Again, no help. Try M in all three slots.

A3OOD3L ASS IN THE M IS HO. HOSTEL IN THE
DE + ILS SEAT 1ORTM ONE DE3REES AND THIRTEEN
M IN ?TES NORTH EAST AND MM NORTH MAIN
MR AN – HSE + ENTHLIMM EAST SIDE SHOOT 1ROM
THE LE1TEMEO1 THE DEATHS HEAD A MEE LINE
1ROM THE TREE THRO?3H THE SHOT 1I1TM1EE TO ?T

New words when M is substituted for 9. Go back to the
cryptogram with 2 and : intact.

A3OOD3L ASS IN THE 2 IS HO. HOSTEL IN THE
DE + ILS SEAT 1ORT: ONE DE3REES AND THIRTEEN
M IN ?TES NORTH EAST AND 2: NORTH MAIN
2R AN – HSE + ENTHLIM2 EAST SIDE SHOOT 1ROM
THE LE1TE:EO1 THE DEATHS HEAD A 2EE LINE
1ROM THE TREE THRO?3H THE SHOT 1I1T:1EE TO ?T

1ROM appears twice. Run down the alphabet with the 1. Assume the word is *FROM*. Substitute *F* for 1. Also, a combination in the second line, M IN ?TES, could be *MINUTES*. Substitute *U* for ?. Further, in the second line, rearrange DE3REES to *DEGREES*. Could 3 be the cipher for *G*? Try all three at once.

A GOOD GLASS IN THE 2 IS HO. HOSTEL IN THE DE+ILS SEAT FOR T: ONE DEGREES AND THIRTEEN MINUTES NORTH EAST AND 2: NORTH MAIN 2R AN –HSE+ENTHLIM2 EAST SIDE SHOOT FROM THE LEFT E:E OF THE DEATHS HEAD A 2EE LINE FROM THE TREE THROUGH THE SHOT FIFT: FEET OUT

Run down the alphabet with the combination 2EE. It must be *BEE*. Try *B* for 2.

A GOOD GLASS IN THE B IS HO. HOSTEL IN THE DE+ILS SEAT FOR T: ONE DEGREES AND THIRTEEN MINUTES NORTH EAST AND B: NORTH MAIN BRAN –H SE+ENTH LIMB EAST SIDE SHOOT FROM THE LEFT E:E OF THE DEATHS HEAD A BEE LINE FROM THE TREE THROUGH THE SHOT FIFT: FEET OUT

Run down the alphabet with the combination E:E. It must be *EYE*. Try *Y* for :. Do the same for DE+ILS. It must be *DEVILS*. Try *V* for +.

A GOOD GLASS IN THE B IS HO. HOSTEL IN THE DEVILS SEAT FORTY ONE DEGREES AND THIRTEEN MINUTES NORTH EAST AND BY NORTH MAIN BRAN –H SEVENTH LIMB EAST SIDE SHOOT FROM THE LEFT EYE OF THE DEATHS HEAD A BEE LINE FROM THE TREE THROUGH THE SHOT FIFTY FEET OUT

Run down the alphabet with the combination BRAN-H. It must be *BRANCH*. Bring the combination B IS HO. together. Run down the alphabet. Since it appears to be the name of a hostel, it is a proper name and could be anything actually. *BISHOP* is the only English word.

P: A GOOD GLASS IN THE BISHOP HOSTEL IN THE DEVILS SEAT FORTY ONE DEGREES AND THIRTEEN MINUTES NORTH EAST AND BY NORTH MAIN BRANCH SEVENTH LIMB EAST SIDE SHOOT FROM THE LEFT EYE OF THE DEATHS HEAD A BEE LINE FROM THE TREE THROUGH THE SHOT FIFTY FEET OUT

```
P: A G O O D G L A S S  I N T H E B I  S H O P H O S
C: 5 3 ! ! / 3 0 5 ) ) 6 * ; 4 8 2 6 ) 4 ! . 4 ! )
   T E L I N T H E D E V I L S S E A T F O R T Y O
   ; 8 0 6 * ; 4 8 / 8 + 6 0 ) ) 8 5 ; 1 ! ( ; : !
   N E D E G R E E S A N D T H I R T E E N M I N U
   * 8 / 8 3 ( 8 8 ) 5 * / ; 4 6 ( ; 8 8 * 9 6 * ?
   T E S N O R T H E A S T A N D B Y N O R T H M A
   ; 8 ) * ! ( ; 4 8 5 ) ; 5 * / 2 : * ! ( ; 4 9 5
   I N B R A N C H S E V E N T H L I M B E A S T S
   6 * 2 ( 5 * – 4 ) 8 + 8 * ; 4 0 6 9 2 8 5 ) ; )
   I D E S H O O T F R O M T H E L E F T E Y E O F
   6 / 8 ) 4 ! ! ; 1 ( ! 9 ; 4 8 0 8 1 ; 8 : 8 ! 1
   T H E D E A T H S H E A D A B E E L I N E F R O
   ; 4 8 / 8 5 ; 4 ) 4 8 5 / 5 2 8 8 0 6 * 8 1 ( !
   M T H E T R E E T H R O U G H T H E S H O T F I
   9 ; 4 8 ; ( 8 8 ; 4 ( ! ? 3 4 ; 4 8 ) 4 ! ; 1 6
   F T Y F E E T O U T
   1 ; : 1 8 8 ; ! ?
```

This cipher was used by Edgar Allen Poe in the tale "The Gold Bug." Poe was profoundly interested in symbols, ciphers, and secret writing. During his life he was recognized as an expert on cryptography.

This cipher falls into the random monoalphabetical substitution category. It has no key. A famous cipher, but so easily cracked that this type is of little use in a practical situation. Since there is no key, the intended receiver of this message would have to know the random symbols and the order of the substitutes. An adversary could easily go through the process, employ a system such as we used, and soon be able to read the message.

You should also be able to see, from working with this system, that the use of nonalphabetical symbols does not enhance the security of the system. The cryptanalyst is completely indifferent to the use of Arabic, English, French, Greek, typographer's symbols, doodles, or any other symbols you could devise.

SHIFT SUBSTITUTION SYSTEM

As soon as writing was invented, it was obvious to the powerful political and military establishment that messages, orders, and instructions could be more accurately sent over any reasonable distance. No longer did the principals have to constantly tour the conquered lands to obtain and convey information one on one, or rely upon the memory of the messenger to relay the message verbally. A runner could be dispatched from Rome with a secret order to be carried out anywhere in the Empire. At first, so few people could read and write that the symbols used were as good as a carefully enciphered secret message. However, within a short time it became obvious to the power structure that others could read their messages as well as the intended receiver. Anyone who held even temporary possession of the message and the ability to read the language could understand the message—it was not secret.

Julius Caesar—eventually a victim of treachery within his own ranks—ruled successfully for years with the aid of a simple cipher system that is the basis for many systems still in use today. Here is an example:

Basic Substitution Systems

```
P:      a b c d e f g h i j k l m n o p q r s t u v w x y z
C: A B C D E F G H I J K L M N O P Q R S T U V W X Y Z
```

The plaintext alphabet is written from left to right. Below it, we will shift another alphabetical grouping three places to the left. A, B, and C in the cipher text are left in midair so they are simply tacked on at the opposite end as shown.

```
P: a b c d e f g h i j k l m n o p q r s t u v w x y z
C: D E F G H I J K L M N O P Q R S T U V W X Y Z A B C
```

The sender and recipient must know the direction and amount of shift used—that would be the key. In this case it would be three to the left. Here is a message that is formed using this simple substitution cipher:

```
P: s t o p   i c e b o x
C: V W R S    L F H E R A
```

   Throughout history—both in fiction and in real life—the most often-used cipher systems have been similar to the one used above; that is, each letter of the plaintext alphabet has only one substitute. Within this basic system, countless variations have been devised. You can think of some. Simply reverse pairs of letters in the cipher alphabet as below:

```
P: a b c d e f g h i j k l m n o p q r s t u v w x y z
C: E D G F I H K J M L O N Q P S R U T W V Y X A Z C B
```

Or start at one end with the shifted D, skip every other space, then fill in on the way back till you have a complete cipher alphabet.

```
P: a b c d e f g h i j k l m n o p q r s t u v w x y z
C: D C E B F A G Z H Y I X J W K V L U M T N S O R P Q
```

Here is another form, this time using a keyword. A keyword is chosen that does not have any repeating letters. The keyword is *BROWN*.

P: a b c d e f g h i j k l m n o p q r s t u v w x y z
C: B R O W N A C D E F G H I J K L M P Q S T U V X Y Z

Below is an important concept, a keyword system that allows for easy mixing of the cipher alphabet. Study this carefully. The keyword *TEXAS* will be used.

```
T E X A S
B C D F G
H I J K L
M N O P Q
R U V W Y
Z
```

Written in columns below TEXAS is the balance of the alphabet, in order, leaving out the letters of the keyword. Then take off the columns in some predetermined order such as first column down, then second column down, then third column down, and so forth.

P: a b c d e f g h i j k l m n o p q r s t u v w x y z
C: T B H M R Z E C I N U X D J O V A F K P W S G L Q Y

As you can see, none of the original order remains in the cipher alphabet. Endless variations have been devised for rearranging the cipher alphabet according to some pre-determined method. It would seem that a system such as this could be used as a practical cipher, but these mono-alphabetic systems are subject to the same treatment we gave the "Goldbug" cipher. They are risky and do not deserve much consideration for use as a practical field cipher.

As we saw in the solution of the "Goldbug" cipher, the favorite tool of the cryptanalyst in the solution of substitution ciphers is the frequency count. (You will find a frequency table in the Appendix.) It shows the frequency of appearance of all of the letters of the alphabet counted from large samples of normal text. It is easy to see that e is the most common letter used, followed by t, a, o, n, i, and so forth. The cryptanalyst makes a count of the symbols in the cryptogram and compares them to the table. Theoretically, if it is a simple substitution cipher, the count of the symbols used should have parallels in the table.

Substitution cryptograms can be devised that distort the frequency, but in the long run, the cryptanalyst will win out because he will simply shift, for the focus of his analysis, to the least frequently used symbols. For example, if he finds a missing letter in the cryptogram, he will assume it to be a consonant, and it will likely be one of the consonants at the bottom of the table such as z, j, x, or q. (The symbols for j, k, q, w, x, and z were missing from the "Goldbug" cipher.)

With just several letters to guide him, a professional cryptanalyst can break the system. Look at the other tables in the Appendix. You may wish to study them in order to avoid using the high-frequency letters as well as some of the common digrams and trigrams when enciphering plaintext. The th in the, then, they, those, and so forth, is a dead giveaway. The wise cryptographer will attempt in every case to avoid these and to distort the message so that it will present more difficulty to the cryptanalyst.

MULTIPLE SUBSTITUTION SYSTEMS

Let us examine some of the systems that use multiple substitution in a specialized sense. Consider an unusual system that is within the simple substitution category. An easily memorized phrase is used for the cipher alphabet. It is used like a key.

P: a b c d e f g h i j k l m n o p q r s t u v w x y z
C: F O U R S C O R E A N D S E V E N Y E A R S A G O O

Note that the cipher alphabet has repetitions. O can substitute for b, g, y, and z.

TABLE SYSTEMS

These systems are especially important since they can be recreated easily by the use of a small table. Here is an example:

	A	B	C	D	E
	F	G	H	I	J
K L M	a	b	c	d	e
N O P	f	g	h	i	j
Q R S	k	l	m	n	o
T U V	p	q	r	s	t
W X Y Z	u	v	w	x	*

* both y and z

The plaintext *a* can be represented by KA, KF, LA, LF, MA, MF, AK, AL, AM, FK, FL, and FM. The *r* could be represented by TC, TH, UC, UH, VC, VH, CT, CU, CV, HT, HU, and HV. This system, while still vulnerable to the codebreaker, is a considerable improvement over the simpler types. This cipher is almost good enough to use as a practical field system, but not quite. We will improve it substantially later.

CHAPTER 5

Polyalphabetic Cipher Systems

■

Carefully constructed transposition and simple-substitution ciphers can be employed in the field with a relative degree of security—for a short period of time—using short messages containing appropriate nulls. They are easy to use, and if the time value of the message is short, they may well be all you need. For messages that must be more secure, however, we must make use of more complex systems such as the polyalphabetical ciphers.

PERIODIC POLYALPHABETICAL CIPHERS

The generic alphabetical table below forms the basis for understanding and using all of the many varieties of polyalphabetical systems.

```
   a b c d e f g h i j k l m n o p q r s t u v w x y z

A  A B C D E F G H I J K L M N O P Q R S T U V W X Y Z
B  B C D E F G H I J K L M N O P Q R S T U V W X Y Z A
C  C D E F G H I J K L M N O P Q R S T U V W X Y Z A B
D  D E F G H I J K L M N O P Q R S T U V W X Y Z A B C
E  E F G H I J K L M N O P Q R S T U V W X Y Z A B C D
F  F G H I J K L M N O P Q R S T U V W X Y Z A B C D E
G  G H I J K L M N O P Q R S T U V W X Y Z A B C D E F
H  H I J K L M N O P Q R S T U V W X Y Z A B C D E F G
I  I J K L M N O P Q R S T U V W X Y Z A B C D E F G H
J  J K L M N O P Q R S T U V W X Y Z A B C D E F G H I
K  K L M N O P Q R S T U V W X Y Z A B C D E F G H I J
L  L M N O P Q R S T U V W X Y Z A B C D E F G H I J K
M  M N O P Q R S T U V W X Y Z A B C D E F G H I J K L
N  N O P Q R S T U V W X Y Z A B C D E F G H I J K L M
O  O P Q R S T U V W X Y Z A B C D E F G H I J K L M N
P  P Q R S T U V W X Y Z A B C D E F G H I J K L M N O
Q  Q R S T U V W X Y Z A B C D E F G H I J K L M N O P
R  R S T U V W X Y Z A B C D E F G H I J K L M N O P Q
S  S T U V W X Y Z A B C D E F G H I J K L M N O P Q R
T  T U V W X Y Z A B C D E F G H I J K L M N O P Q R S
U  U V W X Y Z A B C D E F G H I J K L M N O P Q R S T
V  V W X Y Z A B C D E F G H I J K L M N O P Q R S T U
W  W X Y Z A B C D E F G H I J K L M N O P Q R S T U V
X  X Y Z A B C D E F G H I J K L M N O P Q R S T U V W
Y  Y Z A B C D E F G H I J K L M N O P Q R S T U V W X
Z  Z A B C D E F G H I J K L M N O P Q R S T U V W X Y
```

A table like this first appeared in about 1580 in French literature. It has been reinvented many times since, and there are countless variations and methods based on this simple table. Here is a demonstration. The keyword *SUNDAY* is written repeatedly over the message.

K: S U N D A Y S U N D A Y S U N D A
P: g r e e n a p p l e i s r e a d y

The keyword has six letters, meaning that we will use six different alphabets from the table. At the end of the sixth alphabet, we will start over using the first alphabet. We can say that this cipher has a period of six. Obviously, longer keywords are more secure.

In the case of the first letter of the message, look in the S alphabet, that is, the row that begins with the letter S. Go across until you are under the first letter of the plaintext message: g. Where the two intersect, the S row and the g column, you will find the first letter of the cipher: Y.

Next use the U alphabet. Look for the cipher under r. It is L.

Next use the N alphabet. Find the cipher R under the e column.

Next use the D alphabet. In this case, e is enciphered as H. Carry out the balance of the substitutions.

K: SUNDA YSUND AYSUN DA   (Keyword repeated)
P: green apple isrea dy
C: YLRHN YHJYH IQJYN GY

Throughout history, this sort of cipher in all its variations has been offered to governments, military organizations, and the general public as an unbreakable cipher. This is far from the truth. Professional cryptanalysts can solve a cipher like the one above with or without the keyword and with or without possession of a probable or known word. The periodic nature of the cipher is its weak point. Consider the use of a keyword of only two letters—the system would use only two alphabets and have a period of two.

What about using a longer keyword? Yes, the longer the better. THEQUICKBROWNFOXJUMPEDOVER would make a good keyword, using twenty-two of the alphabets. The period would be so long that the cryptanalyst might at first think it was infinite. It would take a great deal of time to crack, especially if the message was short, had nulls inserted in the plaintext, and was restricted in its use of high-frequency letters, digrams, trigrams, and words.

Here is another form of the polyalphabetical table.

AB	A B C D E F G H I J K L M N O P Q R S T U V W X Y Z
CD	A B C D E F G H I J K L M O P Q R S T U V W X Y Z N
EF	A B C D E F G H I J K L M P Q R S T U V W X Y Z N O
GH	A B C D E F G H I J K L M Q R S T U V W X Y Z N O P
IJ	A B C D E F G H I J K L M R S T U V W X Y Z N O P Q
KL	A B C D E F G H I J K L M S T U V W X Y Z N O P Q R
MN	A B C D E F G H I J K L M T U V W X Y Z N O P Q R S
OP	A B C D E F G H I J K L M U V W X Y Z N O P Q R S T
QR	A B C D E F G H I J K L M V W X Y Z N O P Q R S T U
ST	A B C D E F G H I J K L M W X Y Z N O P Q R S T U V
UV	A B C D E F G H I J K L M X Y Z N O P Q R S T U V W
WX	A B C D E F G H I J K L M Y Z N O P Q R S T U V W X
YZ	A B C D E F G H I J K L M Z N O P Q R S T U V W X Y

Polyalphabetic Cipher Systems

It has been used in many ways. This is a common method.

```
K: B O S T O  N B R A V  E S B O S  T O N B R  A V
P: b e g i n  o m a h a  t h r e e  h u n d r  e d
```

The letters in the keyword select one of thirteen cipher alphabets. Each cipher alphabet can be selected by either of two keyletters. In this case, B, the first letter in the keyword, calls for the use of the first (AB) alphabet.

AB	A B C D E F G H I J K L M
	N O P Q R S T U V W X Y Z

Find the plaintext letter b, then the cipher is below or above it. In this case, O.

The next keyletter, O, causes us to use the OP alphabet. Look for it in the complete table. The plaintext e is found and the Y below it is used as the cipher.

The next keyletter, S, selects the ST alphabet. The plaintext letter g is found and the P below it is the cipher.

The complete cryptogram:

```
K: B O S T O  N B R A V  E S B O S  T O N B R  A V
P: b e g i n  o m a h a  t h r e e  h u n d r  e d
C: O Y P R G  I Z V U X  E Q E Y N  Q A H Q J  R N
```

Notice that in each of the alphabets used in this system, the top half is the same. In the interest of portability and ease of reconstruction in the field, we could shorten the table by using this only once, at the top.

	A B C D E F G H I J K L M
AB	N O P Q R S T U V W X Y Z
CD	O P Q R S T U V W X Y Z N
EF	P Q R S T U V W X Y Z N O
GH	Q R S T U V W X Y Z N O P
IJ	R S T U V W X Y Z N O P Q
KL	S T U V W X Y Z N O P Q R
MN	T U V W X Y Z N O P Q R S
OP	U V W X Y Z N O P Q R S T
QR	V W X Y Z N O P Q R S T U
ST	W X Y Z N O P Q R S T U V
UV	X Y Z N O P Q R S T U V W
WX	Y Z N O P Q R S T U V W X
YZ	Z N O P Q R S T U V W X Y

NONPERIODIC AUTOENCIPHERMENT SYSTEMS

On page 53 is another version of the first table used in this chapter.

```
A B C D E F G H I J K L M N O P Q R S T U V W X Y Z
B C D E F G H I J K L M N O P Q R S T U V W X Y Z A
C D E F G H I J K L M N O P Q R S T U V W X Y Z A B
D E F G H I J K L M N O P Q R S T U V W X Y Z A B C
E F G H I J K L M N O P Q R S T U V W X Y Z A B C D
F G H I J K L M N O P Q R S T U V W X Y Z A B C D E
G H I J K L M N O P Q R S T U V W X Y Z A B C D E F
H I J K L M N O P Q R S T U V W X Y Z A B C D E F G
I J K L M N O P Q R S T U V W X Y Z A B C D E F G H
J K L M N O P Q R S T U V W X Y Z A B C D E F G H I
K L M N O P Q R S T U V W X Y Z A B C D E F G H I J
L M N O P Q R S T U V W X Y Z A B C D E F G H I J K
M N O P Q R S T U V W X Y Z A B C D E F G H I J K L
N O P Q R S T U V W X Y Z A B C D E F G H I J K L M
O P Q R S T U V W X Y Z A B C D E F G H I J K L M N
P Q R S T U V W X Y Z A B C D E F G H I J K L M N O
Q R S T U V W X Y Z A B C D E F G H I J K L M N O P
R S T U V W X Y Z A B C D E F G H I J K L M N O P Q
S T U V W X Y Z A B C D E F G H I J K L M N O P Q R
T U V W X Y Z A B C D E F G H I J K L M N O P Q R S
U V W X Y Z A B C D E F G H I J K L M N O P Q R S T
V W X Y Z A B C D E F G H I J K L M N O P Q R S T U
W X Y Z A B C D E F G H I J K L M N O P Q R S T U V
X Y Z A B C D E F G H I J K L M N O P Q R S T U V W
Y Z A B C D E F G H I J K L M N O P Q R S T U V W X
Z A B C D E F G H I J K L M N O P Q R S T U V W X Y
```

It has been stripped of its labeling letters along the top row and the left column. Essentially it is still the same since the top row and left column in the body of the table are replicas of the label row and column. We will employ this table to illustrate an easy method of obtaining a nonperiodic cipher, yet retain the desirable feature of using a short keyword. The keyword is PARIS. As usual, the keyword is placed above the plaintext.

K: P A R I S
P: r e q u i   r e c o n   f i r m a   t i o n b
y r e d f   o x n n n   (Plaintext with nulls nnn)

Instead of repeating the keyword, however, we will follow it by the plaintext.

K: P A R I S   R E Q U I   R E C O N   F I R M A
P: r e q u i   r e c o n   f i r m a   t i o n b
C: G E H C A   I I S I V   WM T A N   Y Q F Z B
   T I O N B   Y R E D F
   y r e d f   o x n n n
   R G S Q G   M O R Q S

Obviously, as soon as the intended receiver decrypts the material the same length as the keyword, he knows to begin using the plaintext *requi* he has obtained as the continuing keyword material.

There is, of course, no limit as to the length of such a cryptogram, and the encipherment may take place as the message is entered into a computer, then immediately be sent out. Continuous messages such as this, however, are not at all as secure as they may seem at first glance. Once the cryptanalyst discovers the shift being used, the problem of cracking the cipher is similar to other keyword systems.

Some innovators would use the cryptogram itself as the extension of the keyword. Why might this be the wrong thing to do?

MIXED ALPHABET

```
  T U E S D A Y B C F G H I J K L M N O P Q R V W X Z
A ABCDEFGHI J KLMNOPQRS TUVWXYZ
B BCDEFGHI J KLMNOPQRS TUVWXYZA
C CDEFGHI J KLMNOPQRS TUVWXYZAB
D DEFGHI J KLMNOPQRS TUVWXYZABC
E EFGHI J KLMNOPQRS TUVWXYZABCD
F FGHI J KLMNOPQRS TUVWXYZABCDE
G GHI J KLMNOPQRS TUVWXYZABCDEF
H HI J KLMNOPQRS TUVWXYZABCDEFG
I I J KLMNOPQRS TUVWXYZABCDEFGH
J J KLMNOPQRS TUVWXYZABCDEFGHI
K KLMNOPQRS TUVWXYZABCDEFGHI J
L LMNOPQRS TUVWXYZABCDEFGHI JK
M MNOPQRS TUVWXYZABCDEFGHI JKL
N NOPQRS TUVWXYZABCDEFGHI JKLM
O OPQRS TUVWXYZABCDEFGHI JKLMN
P PQRS TUVWXYZABCDEFGHI JKLMNO
Q QRS TUVWXYZABCDEFGHI JKLMNOP
R RS TUVWXYZABCDEFGHI JKLMNOPQ
S S TUVWXYZABCDEFGHI JKLMNOPQR
T TUVWXYZABCDEFGHI JKLMNOPQRS
U UVWXYZABCDEFGHI JKLMNOPQRST
V VWXYZABCDEFGHI JKLMNOPQRSTU
W WXYZABCDEFGHI JKLMNOPQRSTUV
X XYZABCDEFGHI JKLMNOPQRSTUVW
Y YZABCDEFGHI JKLMNOPQRSTUVWX
Z ZABCDEFGHI JKLMNOPQRSTUVWXY
```

Look at the top row. We have altered it to read:

T U E S D A Y B C F G H I J K L M N O P Q R V W X Z

We will use *TUESDAY* now in two ways. It will be the keyword for the encipherment process and it will also be used to mix the alphabets out of regular order. It would still be fairly easy, however, for this table to be set up from scratch and used in the field by anyone who knows the keyword. The mixing of the alphabets adds another level of complication for the would-be codebreaker.

Here is the plaintext message: Report to snowman Mosul immediate.

```
K: T U E S D   A Y T U E   S D A Y T   U E S D A
P: r e p o r   t t o s n   o w m a n   m o s u l
C: O W X K Y   A Y L X V   K A Q D K   K W V E P
   Y T U E S   D A Y T U
   i m m e d   i a t e x
   K J K G W   P F Y V S
```

In the polyalphabetical ciphertext, such as this one, you will see a redistribution of the normal frequency. Note that little-used letters such as *X*, *Y*, *J*, and *V* show up often, even in short messages.

The next logical step would be to add a keyword to the vertical alphabet. We could use the same keyword or another. In any case, it would again add complexity with little added difficulty for the user. Care should be taken to choose keywords that can be committed to memory easily and can logically be placed in proper order. Under stress, even those with nerves of steel and photographic memory have been known to go blank.

The added keyword *PARIS* is placed in the left column. As usual, all these systems can vary; the exact form being selected by the users. Here is a table incorporating these keywords and the resulting mixed alphabets:

```
  T U E S D A Y B C F G H I J K L M N O P Q R V W X Z
P ABCDEFGHI J KLMNOPQRS TUVWXYZ
A BCDEFGHI J KLMNOPQRS TUVWXYZA
R CDEFGHI J KLMNOPQRS TUVWXYZAB
I DEFGHI J KLMNOPQRS TUVWXYZABC
S EFGHI J KLMNOPQRS TUVWXYZABCD
B FGHI J KLMNOPQRS TUVWXYZABCDE
C GHI J KLMNOPQRS TUVWXYZABCDEF
D HI J KLMNOPQRS TUVWXYZABCDEFG
E I J KLMNOPQRS TUVWXYZABCDEFGH
F J KLMNOPQRS TUVWXYZABCDEFGHI
G KLMNOPQRS TUVWXYZABCDEFGHI J
H LMNOPQRS TUVWXYZABCDEFGHI J K
J MNOPQRS TUVWXYZABCDEFGHI J KL
K NOPQRS TUVWXYZABCDEFGHI J KLM
L OPQRS TUVWXYZABCDEFGHI J KLMN
M PQRS TUVWXYZABCDEFGHI J KLMNO
N QRS TUVWXYZABCDEFGHI J KLMNOP
O RS TUVWXYZABCDEFGHI J KLMNOPQ
Q S TUVWXYZABCDEFGHI J KLMNOPQR
T TUVWXYZABCDEFGHI J KLMNOPQRS
U UVWXYZABCDEFGHI J KLMNOPQRST
V VWXYZABCDEFGHI J KLMNOPQRSTU
W WXYZABCDEFGHI J KLMNOPQRSTUV
X XYZABCDEFGHI J KLMNOPQRSTUVW
Y YZABCDEFGHI J KLMNOPQRSTUVWX
Z ZABCDEFGHI J KLMNOPQRSTUVWXY
```

ONE-TIME SYSTEM

In the most critical security cases, many organizations have relied on one of the several varieties of the one-time method. The "one-time" usually refers to the structure and usage of the keyword. Let us suppose that a field agent, posing as a newspaper reporter, carries with him, among his ordinary papers and possessions, some form of unobtrusive written material: a file marked "Research" containing clippings relative to the material he is currently covering. It has been prearranged that he will use the clippings in a certain order, that the encipherment will begin with the twentieth character, and that only alphabetical characters will be used. The intended receiver, of course, has copies of the same clippings. Here is the first clipping:

LOUISIANA BARGE CORP. TO CONSIDER
NEW BID BY BERTRAND OIL GROUP
Louisiana Barge said today it had appointed a committee to study the new proposal from Bertrand Corp. and indicated that the stock swap of the remaining 23% of the Louisiana shares had been settled. The swap is said by observers to have a value of about $230 million. The spokesman for Bertrand...

The plaintext is written under the selected characters from the clipping.

K: O C O N S I D E R N E W B I D B Y B E R
P: r e b e l s h a v e t o m c a t r e q u
T R A N D O I L G R O U P L O U I S I A N A B A R
i r e d o l l a r s o n e h u n d r e d t h o u f
G E S A I D T O D A Y I T H A D A P P O I N T E D
o r r e l e a s e a d v i s e c o m p l i a n c e
A C O M M I T T E E T O S T U D Y T H E N E W P R
c r e d i t t o s w i s s a c c o u n t m a d r i
O P O S A L F R O M B E R T R A N D C O R P A N D
d c o b a n k o n e f i v e n i n e z e r o s e v
I N D I C A T E D T H A T T H
e n c o n t a c t j a e g e r

The agent merges the two sets of characters by using the simple repeated alphabet below:

ABCDEFGHIJKLMNOPQRSTUVWXYZABCDEFGHIJKL
MNOPQRSTUVWXYZ

Placing his pencil on the *O*, the first letter in the work keytext, he counts off the characters until he arrives at the *r*, the first letter in the plaintext. The count is three. Next, he starts at any agreed-upon point in the special alphabet and counts off the three. Let's say that the agreed-upon point is *F*. He counts three, finds *I*, writes *I* for the first cipher-character.

```
K: OCONS  I DERN  EWB I D  BYB ER  TRAND...
P: r e b e l  s h a v e  t omc a  t r e q u  i r e d o...
C: I
```

Next, he counts from *C* to *e*. It is two. He counts from *F* again to find the ciphercharacter. It is *H*. He puts the *H* under the *C* and the *e*.

```
K: OCONS  I DERN  EWB I D  BYB ER  TRAND...
P: r e b e l  s h a v e  t omc a  t r e q u  i r e d o...
C: I H
```

Now he has the pair *O* and *b*. The count from *O* to *b* is 13. Counting 13 from *F* again to obtain the ciphercharacter, he finds *S*.

```
K: OCONS  I DERN  EWB I D  BYB ER  TRAND...
P: r e b e l  s h a v e  t omc a  t r e q u  i r e d o...
C: I HS
```

The next set is *N* to *e*. The count is 17. *F* plus 17 is *W*, the fourth ciphercharacter.

```
K: OCONS  I DERN  EWB I D  BYB ER  TRAND...
P: r e b e l  s h a v e  t omc a  t r e q u  i r e d o...
C: I HSW
```

This simple scheme is continued until the plaintext is completely enciphered. The first five groups are shown on the next page.

K: OCONS IDERN EWBID BYBER TRAND...
P: r e b e l s h a v e t o m c a t r e q u i r e d o...
C: IHSWY PJBJF UXQZC XYIRI VFJVQ...

This is an unbreakable system. It can be used under difficult circumstances with no more than pencil, paper, and the required keytext.

ONE-TIME USING A CODE MACHINE OR SLIDE

A code machine or slide could be used to do most of the mechanical work and reduce errors. A copy of the special repeated alphabet is needed on the edges of two pieces of paper.
First, align all characters to check for errors.

(1) ABCDEFGHI J KLMNOPQRS TUVWXYZABCD
(2) ABCDEFGHI J KLMNOPQRS TUVWXYZABCD
EFGHI JKLMNOPQRS TUVWXYZ
EFGHI JKLMNOPQRS TUVWXYZ

To encipher the first group of the cryptogram we just examined, set O (1) over R (2) and read I (2) below the selected index, F (1).

(1)   ABCDEFGHI J KLMNOPQRS TUVWXYZA
(2) ABCDEFGHI J KLMNOPQRS TUVWXYZABCD
BCDEFGHI J KLMNOPQRS TUVWXYZ
EFGHI JKLMNOPQRS TUVWXYZ

Now set the C over E and read H under the F.

(1)   ABCDEFGHI J KLMNOPQRS TUVWXYZAB
(2) ABCDEFGHI J KLMNOPQRS TUVWXYZABCD
CDEFGHI J KLMNOPQRS TUVWXYZ
EFGHI JKLMNOPQRS TUVWXYZ

Polyalphabetic Cipher Systems

Now set *O* over *B*; read *S* under *F.*

(1) A B C D E F G H I J K L M N O P Q R S T U V W X Y Z A B C D
(2)   A B C D E F G H I J K L M N O P Q
E F G H I J K L M N O P Q R S T U V W X Y Z
R S T U V W X Y Z A B C D E F G H I J K L M N O P Q R S T U V W

Set *N* over *E*; read *W* below *F.*

(1) A B C D E F G H I J K L M N O P Q R S T U V W X Y Z A B C D
(2)   A B C D E F G H I J K L M N O P Q R S T U
E F G H I J K L M N O P Q R S T U V W X Y Z
V W X Y Z A B C D E F G H I J K L M N O P Q R S T U V W X Y Z

Set *S* over *L*; read *Y* under *F.*

(1) A B C D E F G H I J K L M N O P Q R S T U V W X Y Z A B C D
(2)   A B C D E F G H I J K L M N O P Q R S T U V W
E F G H I J K L M N O P Q R S T U V W X Y Z
X Y Z A B C D E F G H I J K L M N O P Q R S T U V W X Y Z

Note that the decipherment process is the reverse of the encipherment. In order to decipher the *Y* just obtained above, set the ciphercharacter *Y* under the index *F*, read the plaintext under the keycharacter *S*. It is *L*.

ONE-TIME USING A TABLE

Now we will adapt this cipher to work using a table similar to the generic table at the first of the chapter. Note that the index *F* has been used to start the alphabet in the top row.

```
  F G H I J K L M N O P Q R S T U V W X Y Z A B C D E
A A B C D E F G H I J K L M N O P Q R S T U V W X Y Z
B B C D E F G H I J K L M N O P Q R S T U V W X Y Z A
C C D E F G H I J K L M N O P Q R S T U V W X Y Z A B
D D E F G H I J K L M N O P Q R S T U V W X Y Z A B C
E E F G H I J K L M N O P Q R S T U V W X Y Z A B C D
F F G H I J K L M N O P Q R S T U V W X Y Z A B C D E
G G H I J K L M N O P Q R S T U V W X Y Z A B C D E F
H H I J K L M N O P Q R S T U V W X Y Z A B C D E F G
I I J K L M N O P Q R S T U V W X Y Z A B C D E F G H
J J K L M N O P Q R S T U V W X Y Z A B C D E F G H I
K K L M N O P Q R S T U V W X Y Z A B C D E F G H I J
L L M N O P Q R S T U V W X Y Z A B C D E F G H I J K
M M N O P Q R S T U V W X Y Z A B C D E F G H I J K L
N N O P Q R S T U V W X Y Z A B C D E F G H I J K L M
O O P Q R S T U V W X Y Z A B C D E F G H I J K L M N
P P Q R S T U V W X Y Z A B C D E F G H I J K L M N O
Q Q R S T U V W X Y Z A B C D E F G H I J K L M N O P
R R S T U V W X Y Z A B C D E F G H I J K L M N O P Q
S S T U V W X Y Z A B C D E F G H I J K L M N O P Q R
T T U V W X Y Z A B C D E F G H I J K L M N O P Q R S
U U V W X Y Z A B C D E F G H I J K L M N O P Q R S T
V V W X Y Z A B C D E F G H I J K L M N O P Q R S T U
W W X Y Z A B C D E F G H I J K L M N O P Q R S T U V
X X Y Z A B C D E F G H I J K L M N O P Q R S T U V W
Y Y Z A B C D E F G H I J K L M N O P Q R S T U V W X
Z Z A B C D E F G H I J K L M N O P Q R S T U V W X Y
```

Here is the same ciphertext and keytext we used above:

K: OCONS IDERN EWBID BYBER TRAND...
P: rebel shave tomca trequ iredo...

In the leftmost column, find the keycharacter O. Go across on that row to the plaintext R. Now go up that column to the top row. You will find I, the ciphercharacter.

Polyalphabetic Cipher Systems

To encipher the next character, the keytext dictates the use of the *C* alphabet, so use the *C* row. Go across to *E*. Then go up that column to the top. *H* is the ciphercharacter.

To encipher the plaintext *b*, use the *O* row across to *B*, up to find the ciphercharacter *S*.

```
K: OCONS  IDERN EWBID BYBER TRAND...
P: r e b e l  s h a v e  t o m c a  t r e q u  i r e d o...
C: IHS...
```

Again, decipherment is the reverse of encipherment.

Later, we will employ this valuable one-time system using the state-of-the-art code machine, the computer. And for complete security plus portability, we will develop a one-time keytext using only paper, pencil, and a $4.99 pocket calculator.

CHAPTER 6

Code Machines

In polygram substitution ciphers, several characters are enciphered together. The most commonly used type is the digram form. That is, two characters are replaced, following some scheme, by two substitute characters. Other systems, carrying the idea further, replace trigrams (three characters) by other trigrams. Further, groups could be replaced by groups, and so forth. But most of these systems using character groups larger than digram size quickly become so complex that computers are required to store the large tables and operate the system.

DIGRAM SYSTEM

Below is a generic table that allows the construction of a digram system. It is of little use, however, since in this form a plaintext would be enciphered into itself.

A B C D E F G H I J K L M N O P Q R S T U V W X Y Z

A AA BA CA DA EA FA GA HA IA JA KA LA MANA OA PA QA RA SA TA UA VA WA XA YA ZA

B AB BB CB DB EB FB GB HB IB JB KB LB MB NB OB PB QB RB SB TB UB VB WB XB YB ZB

C AC BC CC DC EC FC GC HC IC JC KC LC MC NC OC PC QC RC SC TC UC VC WC XC YC ZC

D AD BD CD DD ED FD GD HD ID JD KD LD MD ND OD PD QD RD SD TD UD VD WD XD YD ZD

E AE BE CE DE EE FE GE HE IE JE KE LE ME NE OE PE QE RE SE TE UE VE WE XE YE ZE

F AF BF CF DF EF FF GF HF IF JF KF LF MF NF OF PF QF RF SF TF UF VF WF XF YF ZF

G AG BG CG DG EG FG GG HG IG JG KG LG MG NG OG PG QG RG SG TG UG VG WG XG YG ZG

H AH BH CH DH EH FH GH HH IH JH KH LH MH NH OH PH QH RH SH TH UH VH WH XH YH ZH

I AI BI CI DI EI FI GI HI II JI KI LI MI NI OI PI QI RI SI TI UI VI WI XI YI ZI

J AJ BJ CJ DJ EJ FJ GJ HJ IJ JJ KJ LJ MJ NJ OJ PJ QJ RJ SJ TJ UJ VJ WJ XJ YJ ZJ

K AK BK CK DK EK FK GK HK IK JK KK LK MK NK OK PK QK RK SK TK UK VK WK XK YK ZK

L AL BL CL DL EL FL GL HL IL JL KL LL ML NL OL PL QL RL SL TL UL VL WL XL YL ZL

M AM BM CM DM EM FM GM HM IM JM KM LM MM NM OM PM QM RM SM TM UM VM WM XM YM ZM

N AN BN CN DN EN FN GN HN IN JN KN LN MN NN ON PN QN RN SN TN UN VN WN XN YN ZN

O AO BO CO DO EO FO GO HO IO JO KO LO MO NO OO PO QO RO SO TO UO VO WO XO YO ZO

P AP BP CP DP EP FP GP HP IP JP KP LP MP NP OP PP QP RP SP TP UP VP WP XP YP ZP

Q AQ BQ CQ DQ EQ FQ GQ HQ IQ JQ KQ LQ MQ NQ OQ PQ QQ RQ SQ TQ UQ VQ WQ XQ YQ ZQ

R AR BR CR DR ER FR GR HR IR JR KR LR MR NR OR PR QR RR SR TR UR VR WR XR YR ZR

S AS BS CS DS ES FS GS HS IS JS KS LS MS NS OS PS QS RS SS TS US VS WS XS YS ZS

T AT BT CT DT ET FT GT HT IT JT KT LT MT NT OT PT QT RT ST TT UT VT WT XT YT ZT

U AU BU CU DU EU FU GU HU IU JU KU LU MU NU OU PU QU RU SU TU UU VU WU XU YU ZU

V AV BV CV DV EV FV GV HV IV JV KV LV MV NV OV PV QV RV SV TV UV VV WV XV YV ZV

W AW BW CW DW EW FW GW HW IW JW KW LW MW NW OW PW QW RW SW TW UW VW WW XW YW ZW

X AX BX CX DX EX FX GX HX IX JX KX LX MX NX OX PX QX RX SX TX UX VX WX XX YX ZX

Y AY BY CY DY EY FY GY HY IY JY KY LY MY NY OY PY QY RY SY TY UY VY WY XY YY ZY

Z AZ BZ CZ DZ EZ FZ GZ HZ IZ JZ KZ LZ MZ NZ OZ PZ QZ RZ SZ TZ UZ VZ WZ XZ YZ ZZ

We will change the external alphabets, that is, mix them, by using the keywords *JULY* and *EASY*.

J U L Y A B C D E F G H I K M N O P Q R S T V W X Z

	J	U	L	Y	A	B	C	D	E	F	G	H	I	K	M	N	O	P	Q	R	S	T	V	W	X	Z
E	AA	BA	CA	DA	EA	FA	GA	HA	IA	JA	KA	LA	MA	NA	OA	PA	QA	RA	SA	TA	UA	VA	WA	XA	YA	ZA
A	AB	BB	CB	DB	EB	FB	GB	HB	IB	JB	KB	LB	MB	NB	OB	PB	QB	RB	SB	TB	UB	VB	WB	XB	YB	ZB
S	AC	BC	CC	DC	EC	FC	GC	HC	IC	JC	KC	LC	MC	NC	OC	PC	QC	RC	SC	TC	UC	VC	WC	XC	YC	ZC
Y	AD	BD	CD	DD	ED	FD	GD	HD	ID	JD	KD	LD	MD	ND	OD	PD	QD	RD	SD	TD	UD	VD	WD	XD	YD	ZD
B	AE	BE	CE	DE	EE	FE	GE	HE	IE	JE	KE	LE	ME	NE	OE	PE	QE	RE	SE	TE	UE	VE	WE	XE	YE	ZE
C	AF	BF	CF	DF	EF	FF	GF	HF	IF	JF	KF	LF	MF	NF	OF	PF	QF	RF	SF	TF	UF	VF	WF	XF	YF	ZF
D	AG	BG	CG	DG	EG	FG	GG	HG	IG	JG	KG	LG	MG	NG	OG	PG	QG	RG	SG	TG	UG	VG	WG	XG	YG	ZG
F	AH	BH	CH	DH	EH	FH	GH	HH	IH	JH	KH	LH	MH	NH	OH	PH	QH	RH	SH	TH	UH	VH	WH	XH	YH	ZH
G	AI	BI	CI	DI	EI	FI	GI	HI	II	JI	KI	LI	MI	NI	OI	PI	QI	RI	SI	TI	UI	VI	WI	XI	YI	ZI
H	AJ	BJ	CJ	DJ	EJ	FJ	GJ	HJ	IJ	JJ	KJ	LJ	MJ	NJ	OJ	PJ	QJ	RJ	SJ	TJ	UJ	VJ	WJ	XJ	YJ	ZJ
I	AK	BK	CK	DK	EK	FK	GK	HK	IK	JK	KK	LK	MK	NK	OK	PK	QK	RK	SK	TK	UK	VK	WK	XK	YK	ZK
J	AL	BL	CL	DL	EL	FL	GL	HL	IL	JL	KL	LL	ML	NL	OL	PL	QL	RL	SL	TL	UL	VL	WL	XL	YL	ZL
K	AM	BM	CM	DM	EM	FM	GM	HM	IM	JM	KM	LM	MM	NM	OM	PM	QM	RM	SM	TM	UM	VM	WM	XM	YM	ZM
L	AN	BN	CN	DN	EN	FN	GN	HN	IN	JN	KN	LN	MN	NN	ON	PN	QN	RN	SN	TN	UN	VN	WN	XN	YN	ZN
M	AO	BO	CO	DO	EO	FO	GO	HO	IO	JO	KO	LO	MO	NO	OO	PO	QO	RO	SO	TO	UO	VO	WO	XO	YO	ZO
N	AP	BP	CP	DP	EP	FP	GP	HP	IP	JP	KP	LP	MP	NP	OP	PP	QP	RP	SP	TP	UP	VP	WP	XP	YP	ZP
O	AQ	BQ	CQ	DQ	EQ	FQ	GQ	HQ	IQ	JQ	KQ	LQ	MQ	NQ	OQ	PQ	QQ	RQ	SQ	TQ	UQ	VQ	WQ	XQ	YQ	ZQ
P	AR	BR	CR	DR	ER	FR	GR	HR	IR	JR	KR	LR	MR	NR	OR	PR	QR	RR	SR	TR	UR	VR	WR	XR	YR	ZR
Q	AS	BS	CS	DS	ES	FS	GS	HS	IS	JS	KS	LS	MS	NS	OS	PS	QS	RS	SS	TS	US	VS	WS	XS	YS	ZS
R	AT	BT	CT	DT	ET	FT	GT	HT	IT	JT	KT	LT	MT	NT	OT	PT	QT	RT	ST	TT	UT	VT	WT	XT	YT	ZT
T	AU	BU	CU	DU	EU	FU	GU	HU	IU	JU	KU	LU	MU	NU	OU	PU	QU	RU	SU	TU	UU	VU	WU	XU	YU	ZU
U	AV	BV	CV	DV	EV	FV	GV	HV	IV	JV	KV	LV	MV	NV	OV	PV	QV	RV	SV	TV	UV	VV	WV	XV	YV	ZV
V	AW	BW	CW	DW	EW	FW	GW	HW	IW	JW	KW	LW	MW	NW	OW	PW	QW	RW	SW	TW	UW	VW	WW	XW	YW	ZW
W	AX	BX	CX	DX	EX	FX	GX	HX	IX	JX	KX	LX	MX	NX	OX	PX	QX	RX	SX	TX	UX	VX	WX	XX	YX	ZX
X	AY	BY	CY	DY	EY	FY	GY	HY	IY	JY	KY	LY	MY	NY	OY	PY	QY	RY	SY	TY	UY	VY	WY	XY	YY	ZY
Z	AZ	BZ	CZ	DZ	EZ	FZ	GZ	HZ	IZ	JZ	KZ	LZ	MZ	NZ	OZ	PZ	QZ	RZ	SZ	TZ	UZ	VZ	WZ	XZ	YZ	ZZ

Now encipher re, row first, column second. It is IT. Each time the pair re would appear in the plaintext, the substitute IT would be the cipher. This is a form of simple substitution. If the same pair, but in reversed order, appears, er, it will be enciphered as TA. Modern day cryptanalysts would quickly see that certain pairs had higher frequency than others and would make quick work of the system, but ciphers such as this were used for centuries by governments and military organizations, and while they are of little practical use to us today, the student should carefully study the concept since we will use it to great advantage later in a sophisticated, virtually unbreakable fractional system.

SMALL TABLE DIGRAM SYSTEM

```
A B C D E
F G H I J
K L M N O
P Q R S T
U V W X *
```

*Y and Z are
in this space.

In this method, the plaintext pairs are thought of as forming two corners of a rectangle. The cipher pairs are the remaining corners. If no rectangle is formed within the table, or if there is a pair that is a double, other rules apply as explained below.

P: redtop eliminated paris send encore

Arrange in two-character groups.

P: re dt op el im in at ed pa ri ss en de nc or e

The first pair, re, has TC as the alternate corners of its rectangle. It is TC, not CT. The substitute is obtained from the same row and order as the plaintext in the case of diagonals; dt, the second pair, has ES as its alternate; op becomes KT, el becomes BO, and so forth.

P: re dt op el im in at ed pa ri ss en
C: TC ES KT BO HN NS EP AE
de nc or e

In *ed*, and similar cases, a proper rectangle cannot be formed. When this happens, substitution is made to the right when the pair is in the same row. *E* is substituted for *d*. When the members of the pair are in the same column, the substitution is made downward.

Additionally, if one of any pair is in the rightmost column, the substitute is made with the character on the same row but in the leftmost column. Therefore, *e* has *A* as its substitute.

A similar rule applies if one of the pair falls in the bottom row. The substitute is found in the same column but in the top row.

```
P:  r e  d t  op  e l  im  in  at  ed  pa  r i  ss  en
C:  TC  ES  KT  BO  HN  NS  EP  AE  KU  SH  TT
de  nc  or  e
```

When a double is encountered, it follows the "substitute-to-the-right" rule. The *e* at the end could be paired with a null.

Effectively, this is another example of modified simple substitution. Each distinctive plaintext pair is always replaced by the same substitute pair. Obviously, in this form, it is a dangerous system with little security. The concept, however, has merit, and with refinement it can be made into a respectable cipher.

FRACTIONAL METHODS

```
P:  r e  d t  op  e l  im  in  at  ed  pa  r i  ss  en
C:  TC  ES  KT  BO  HN  NS  EP  AE  KU  SH  TT  DO
    T   E   K   B   H   N   E   A   K   S   T   D
    C   S   T   O   N   S   P   E   U   H   T   O
de  nc  or  e
EA  MD  MT  A
E   M   M   A
    A   D   T
C:  TEKBH  NEAKS  TDEMM  ACSTO  NSPEU  HTOAD  T
```

On the previous page, the cipher pairs are broken apart then combined according to a simple alternate transposition system covered in the first chapters of this manual. This is actually a pseudofractional system. In order to fractionate a character, we must form two characters from one character of plaintext. This would be one way.

```
  A B C D E F G H I J K L M N O P Q R S T U V W X Y Z

A A B C D E F G H I J K L M N O P Q R S T U V W X Y Z
B B C D E F G H I J K L M N O P Q R S T U V W X Y Z A
C C D E F G H I J K L M N O P Q R S T U V W X Y Z A B
D D E F G H I J K L M N O P Q R S T U V W X Y Z A B C
E E F G H I J K L M N O P Q R S T U V W X Y Z A B C D
F F G H I J K L M N O P Q R S T U V W X Y Z A B C D E
G G H I J K L M N O P Q R S T U V W X Y Z A B C D E F
H H I J K L M N O P Q R S T U V W X Y Z A B C D E F G
I I J K L M N O P Q R S T U V W X Y Z A B C D E F G H
J J K L M N O P Q R S T U V W X Y Z A B C D E F G H I
K K L M N O P Q R S T U V W X Y Z A B C D E F G H I J
L L M N O P Q R S T U V W X Y Z A B C D E F G H I J K
M M N O P Q R S T U V W X Y Z A B C D E F G H I J K L
N N O P Q R S T U V W X Y Z A B C D E F G H I J K L M
O O P Q R S T U V W X Y Z A B C D E F G H I J K L M N
P P Q R S T U V W X Y Z A B C D E F G H I J K L M N O
Q Q R S T U V W X Y Z A B C D E F G H I J K L M N O P
R R S T U V W X Y Z A B C D E F G H I J K L M N O P Q
S S T U V W X Y Z A B C D E F G H I J K L M N O P Q R
T T U V W X Y Z A B C D E F G H I J K L M N O P Q R S
U U V W X Y Z A B C D E F G H I J K L M N O P Q R S T
V V W X Y Z A B C D E F G H I J K L M N O P Q R S T U
W W X Y Z A B C D E F G H I J K L M N O P Q R S T U V
X X Y Z A B C D E F G H I J K L M N O P Q R S T U V W
Y Y Z A B C D E F G H I J K L M N O P Q R S T U V W X
Z Z A B C D E F G H I J K L M N O P Q R S T U V W X Y
```

P: amsterdam wants new deal

Find any *a* in the body of the table. Let us say that the *a* selected is at the intersection of the coordinates of *MO*. Or it could be *OM*, since it points to *a* as well. In this case, we will substitute *MO* for *a*. Now, if we separate *MO* into *M* and *O*, *M* is meaningless in itself, and the only thing that can be said of *M* is that it is a part, or fraction, of *a, b, c, d . . . x, y*, or *z*, since all members of the alphabet are on the *M* row. Likewise the *O* can represent a part of any character in the alphabet. *M* and *O* are only specifically meaningful when they are together.

```
P:  a   m   s   t   e   r   d   a   m   w   a   n   t   s
C: MO HF AS SB XH KH NQ TH ZN  SE ...
    n   e   w   d   e   a   l
```

(This would have to be completed to be correct.)
Now transpose.

```
  M  H  A  S  X  K  N  T  Z  S
   O  F  S  B  H  H  Q  H  N  E
C: MHASX KNTZS OFSBH HQHNE ...
```

We see here the beginnings of a good *fractional product cipher*. Even in this form it would make tough chewing for the codebreaker. In the next chapter we will develop a compact version of this system.

SLIDES, GRILLES AND OTHER CODE MACHINES

Any device which mechanically or electronically aids in the encipherment or decipherment of a cryptogram can be termed a "code machine." These machines have progressed from early paper-puzzle contrivances through wooden slides, metal disks, complete typewriter-like machines that select polyalphabetical encipherments, and finally to the modern computer. With the exception of the computer, virtually all of these machines are useless today in any practical system. Excellent paper-and-pencil systems that can be read by manual or computer decipherment are available to the serious user. Extremely portable pocket micromicrocomputers are available that, if properly programmed, can generate virtually unbreakable ciphers rapidly and error free. For these reasons we will cover the older code machines only briefly.

The first slide (similar to the one demonstrated in Chapter 5) appeared in about 1880. Many references are made in the literature to the slides used at the Saint Cyr School of the Military in France at about this time. Countless varieties have been in use, but most are either linear or circular, containing two or three alphabets on movable sliding strips or disks.

In the simplest form, both slides must total seventy-seven alphabetical symbols for the English alphabet. They can be set up on index cards or any other convenient material. These disks and slides can be used for monoalphabetical or polyalphabetical encipherment. As we saw earlier, most table systems can be transformed to operate in a slide or disk arrangement.

GRILLES

Grilles are usually made of some material such as paper or thin metal. Openings are cut in them at intervals that correspond to certain characters' positions in an innocent-looking message. When the grille is placed over the message, it reveals only the enciphered message.

The student will encounter from time to time in literature and in military history the mention of cipher systems that use grilles or grids. These have been used in many ways, but all are based on the idea that openings, notches, or markings are in some way used on a paper card, metal sheet, or other material so that when this device is placed over a jumbled secret message, parts of it will show through the openings or be indexed by the markings so that a message can be enciphered or deciphered.

Some grilles were designed to be turned with each character; some were turned at the completion of each word; others were shaped in triangles or hexagons or circles. A single grille can be reversed or turned upside down and used for several different systems.

Grille systems produce excellent encipherments; however, they do have two drawbacks. The grille must be carried in the user's possession, and if captured, the grille exposes the entire system to the adversary. In fact, all code machines are vulnerable from this angle. If it is lost, then the system is lost. If it is captured by an adversary, the system is compromised and all or most of the messages sent in the past can be deciphered. It is for these reasons that few practical systems today use any type of code machine other than the computer. Obviously, you should not use any device that cannot be destroyed or rendered useless almost instantaneously.

An example of a grille system that many of us remember is the IBM punched card. The IBM card allowed electrical contacts or light beams to pass through holes to activate photocells to decode the message. Just as the IBM card was trimmed on one corner, a grille would be coded in some way—notched—so that its proper orientation could be easily ascertained.

Numerous other machines and small devices have been in use throughout the history of cryptography, but they all have given way to the computer. The computer can do far more in constructing cipher systems than any of the mechanical code machines and can do it much faster with less chance of error. Excellent product ciphers—that is, ciphers formed by a combination of substitution and transposition schemes—can now be constructed that were impractical in earlier times because of time requirements and the inevitable human error introduced into complex systems. Programs for computers have been devised that trap errors, do self-checking, do mathematical operations, and easily convert symbols to numbers and back again, giving the encrypter a virtually unlimited horizon for the development of new and better systems.

CHAPTER 7

Codemaster System

■

Literally thousands of codes have been used by the military, governments, commercial businesses, clubs, and other organizations. When first written and put into use, they are secure because they do not ordinarily rely on a cipher system that could be solved by the cryptanalyst. They are much like a random, mixed-alphabet cipher. In most cases, the symbols substituted for the plaintext words are never changed, and as decoded messages fall into public hands, the security value becomes nil. The users rely on the fact that only those who possess a copy of the codebook, listing the substitutes, can easily decode the material.

Many of these codes have been built on empirical systems that have been added to over the years. All of them can be decoded if you are willing to monitor enough messages, determine the basic activity of the organization using the code, and keep records of your intercepts.

Here is a portion of a commercial code in use today. It goes on for hundreds of pages, listing practically every word in the dictionary.

Code	P:
MABTA	unable
MABTE	busy
MABTO	urgent
MABTU	specification
MABTY	paper
MABUB	below
MABUC	model

The first section has the five-letter code group arranged alphabetically. The second section has the plaintext arranged alphabetically. This is an example of a two-part code.

P:	Code
gold	KLUMA
golden	NJUSW
goldstone	LPOIJ
golf	IOKUJ
gondola	QWSED
gone	TGYDA
good	MNNDQ

Ordinarily, codes are extremely large—so large that they have to be printed and bound into book form and distributed directly to the users. In order to change the code, a new codebook must be prepared and placed in the hands of the users.

In some cases, the code is not intended to be absolutely secret; the users simply wish to exclude the casual eavesdropper invariably present in the chain of ordinary communications.

The "10" code, made popular by Broderick Crawford in the early television series "Highway Patrol," is in this category. It is an excellent code used by many police groups. It saves time, provides a certain degree of secrecy, and reduces errors. These are the main attributes of any good code system. It is sometimes referred to as the Police Code or the Police Ten

Codemaster System

Code. Many organizations have adapted portions of this system. Here is the original version:

10–0	Caution
10–1	Do not understand/unable to copy/change your location
10–2	Signal read/able to copy
10–3	Stop transmitting
10–4	Acknowledge
10–5	Relay
10–6	Busy, stand by unless urgent
10–7	Out of service
10–8	In service
10–9	Repeat
10–10	Fight in progress
10–11	Dog fight
10–12	Stand by
10–13	Weather
10–14	Prowler
10–15	Civil disturbance
10–16	Domestic problem
10–17	Meet person
10–18	Complete quickly
10–19	Return to...
10–20	Give location
10–21	Use telephone
10–22	Cancel
10–23	Arrived
10–24	Assignment completed
10–25	Report to...
10–27	License information
10–28	Registration information
10–29	Check for wanted
10–30	Illegal use of radio
10–31	Crime in progress
10–32	Person with a gun
10–33	Emergency

10–34	Riot
10–35	Major alert
10–36	Correct time
10–37	Suspicious vehicle
10–41	Beginning tour of duty
10–42	Ending tour of duty
10–43	Information
10–44	Request permission to suspend patrol
10–45	Animal
10–46	Assist
10–47	Emergency road repairs
10–48	Traffic sign trouble
10–49	Traffic light problem
10–50	Accident
10–51	Wrecker needed
10–52	Ambulance needed
10–53	Road blocked
10–54	Livestock
10–55	Intoxicated driver
10–56	Intoxicated pedestrian
10–57	Hit and run
10–58	Direct traffic
10–59	Escort
10–61	Personnel in area
10–62	Reply
10–63	Need written copy
10–64	New assignment
10–66	Message cancelled
10–67	Clear to read message
10–68	Dispatch information
10–69	Message received
10–70	Fire alarm
10–71	Advise nature of fire
10–72	Report progress of fire
10–73	Smoke
10–74	Negative
10–75	In contact with...

10–76	En route
10–77	ETA (estimated time of arrival)
10–78	Need assistance
10–79	Need coroner
10–84	Are you going to meet. . .
10–85	Delay
10–88	Advise phone number
10–90	Bank alarm
10–94	Racing
10–96	Mental subject
10–98	Prison or jail break
10–99	Indicate wanted or stolen

The citizens band (CB) group adopted the Police Ten Code. Here are the first ten signals as they use it:

10–1	Cannot copy/poor reception
10–2	Can copy/good reception
10–3	Hold it/stop/cease transmission/do not interfere
10–4	I agree/OK/message received
10–5	Relay message
10–6	Busy, stand by
10–7	Out of service, going off air
10–8	In service, OK to call
10–9	Repeat message
10–10	Transmission complete, standing by

Here is another useful code system of long standing. This code has been used primarily by radio operators but has been adapted and used by many other organizations. In practice, the question mark is sent only when the Q signal is meant to be a question. Again, we are listing only the first ten signals to give you an idea of the construction.

QRG	Will you tell me my exact frequency (or that of _____)?

	Your exact frequency (or that of _____) is _____ kHz.
QRH	Does my frequency vary? Your frequency varies.
QRI	How is the tone of my transmission? The tone of your transmission is _____. (1) good (2) variable (3) bad
QRK	What is the intelligibility of my signals (or those of _____)? The intelligibility of your signals (or those of _____) is _____. (1) bad (2) poor (3) fair (4) good (5) excellent
QRL	Are you busy? I am busy (or I am busy with _____). Please do not interfere.
QRM	Is my transmission being interfered with? Your transmission is being interfered with _____. (1) nil (2) slightly (3) moderately (4) severely (5) extremely
QRN	Are you troubled by static? I am troubled by static _____ (1–5 as under QRM).
QRO	Shall I increase power? Increase power
QRP	Shall I decrease power? Decrease power

QRQ   Shall I send faster?
           Send faster (_____ wpm).

Here is another code used by radio people to describe the communication signal:

Readability:

Code	Plaintext
1	Unreadable
2	Barely readable, unsatisfactory
3	Readable with considerable difficulty
4	Readable with practically no difficulty
5	Perfectly readable

Signal strength:

Code	Plaintext
1	Faint signals, barely perceptible
2	Very weak signals
3	Weak signals
4	Fair signals
5	Fairly good signals
6	Good signals
7	Moderately strong signals
8	Strong signals
9	Extremely strong signals

Here is another code—a "phonetic" code for each letter of the alphabet:

A	ALPHA/ALFA	I	INDIA
B	BRAVO	J	JULIETT
C	CHARLIE	K	KILO
D	DELTA	L	LIMA
E	ECHO	M	MIKE
F	FOXTROT	N	NOVEMBER
G	GOLF	O	OSCAR
H	HOTEL	P	PAPA

Q	QUEBEC	V	VICTOR
R	ROMEO	W	WHISKEY
S	SIERRA ·	X	X-RAY
T	TANGO	Y	YANKEE
U	UNIFORM	Z	ZULU

This phonetic system is a useful code, but as you can see, the codewords are not uniform five-letter code groups and contain several confusing spellings. The author proposed in 1978 that the following code, though not perfect, be substituted for the earlier system. All of the references to the phonetic code in this manual use the author's newer version as listed below:

A	ALPHA	N	NOVEL
B	BRAVO	O	OSCAR
C	CARLO	P	PAPER
D	DELTA	Q	QUICK
E	ECHOS	R	ROMEO
F	FOLLY	S	SIGMA
G	GOOSE	T	TANGO
H	HOTEL	U	UNITE
I	INDIA	V	VIRGO
J	JULIE	W	WHITE
K	KILOS	X	XRAYS
L	LIMAS	Y	YALTA
M	MILAN	Z	ZEBRA

BOOK CODES

Book codes and codebooks are not the same. A book code system can be formed easily and quickly using several copies of the same book. Virtually any book can be used if it has the words normally needed by the users in the text. A book of fiction, the Bible, even a dictionary can be used.

A typical system could be operated as follows: the plaintext word would be located in the text. Let us say that it is the eleventh word on page 46 of *Gone with the Wind*. According to a prearranged scheme, the encoder would add 10 to the page number and 5 to the word count. So the code word would be found on page 56, the sixteenth word into the text. The encoder would continue finding the plaintext words anywhere in the book, following the system, until the entire message was encoded. The code words obtained in this way are often superenciphered in an attempt to make the system more difficult for the cryptanalyst.

Innumerable variations, prearranged by the users, can and have been used. The major disadvantage of a book system is that the users must have identical copies of the book in their possession. The entire system cannot be memorized. Further, book codes have been used so often that any book found in the possession of a possible cryptographer or agent is immediately suspect. Most often, the counteroperatives will allow the suspected agent to pass through customs or other checkpoints unmolested. They will then forward a record of any books in his possession to the appropriate intelligence cryptanalysts, and if the agent passes messages in any form based on the use of a book code, they will be more easily broken. In fact, even if they are superenciphered, the cipher may be easier to break because the cryptanalyst has word patterns from the book to help him.

The rule is this: any device or item of any kind that must be kept in the possession of the users detracts from the value of the system. Codebooks and book codes are some of the best examples of these dangerous devices and should be used with extreme caution.

But there are some advantages to a properly designed and implemented code system. In some cases, the message can be shortened. Further, if the codeword system is small, it can be memorized or stored in the memory of a micro-computer or micromicrocomputer. In this form, the code-words, if superenciphered, are indeed more secure.

Only micromicrocomputers that can masquerade as calculators or database holders are generally satisfactory for use in the field. The larger micros can be used in upper echelons and can be programmed to encipher and decipher communications to the micromicros or to paper and pencil systems. Computers used in the field must have one important feature: a reset or panic button that erases instantaneously all programs and data from the memory of the computer. An extremely dangerous item that is sold with some micromicros is the EPROM type ROM. EPROM means Erasable Programmable Read Only Memory. Programs and data are "burned into" EPROMs and are more or less permanent. In order to be erased, they must be exposed to intense ultraviolet light for an extended period of time. This can hardly be done in most field situations. Computers containing EPROM modules can sustain considerable damage, yet the data from the EPROMs can be salvaged. A better alternative, if preprogrammed modules must be used, would be the RAM (Random Access Memory) cartridges that contain a battery. When the battery is removed, all programming and data is instantly and irretrievably lost.

Codewords are used in the open, but often they are superenciphered into the cipher in current use. In any case, it has frequently been a fatal mistake to use a codeword system in superencipherment that is changed infrequently or not at all. Codewords used in this way can first become patterns to the cryptanalyst and eventually be recognized as probable words—the most dangerous words a cryptographer can use. Obviously, if his unit uses the Q signals, the 10 codes, or any of the common codes—in the open (or clear)—a cryptographer should never encipher any of them into the contemporary cipher systems he is employing.

The ideal codeword system for use in the field is small, can be quickly changed, and should be easily convertible to voice or Morse communications. Numerical characters should not be used in the codewords because they are difficult to transmit both by voice and Morse, and the codewords would almost immediately be recognized for what they are.

The *Codemaster* system meets these requirements and, as

Codemaster System

a bonus, can be operated using a keyword system. First we will construct a simple twenty-six word manual system. This can be nothing more than a list of words that is often used by your unit paired with the Codemaster Phonetic Alphabet.

Codeword	Plaintext	Codeword	Plaintext
ALPHA	MISSION	NOVEL	WITH
BRAVO	LOCATED	OSCAR	TAKE
CARLO	CONTACT	PAPER	BRING
DELTA	RETRN	QUICK	GET
ECHOS	SILENCE	ROMEO	SEND
FOLLY	COPY	SIGMA	MEET
GOOSE	START	TANGO	SUPPLY
HOTEL	FINISH	UNITE	LOW
INDIA	MINUTE	VIRGO	HIGH
JULIE	HOUR	WHITE	EAST
KILOS	TODAY	XRAYS	WEST
LIMAS	TO	YALTA	NO
MILAN	AT	ZEBRA	NORTH

The next step is to use a keyword to rearrange the codeword order. Select a word that has no repetitive characters. We will use *EAST.*

Codeword	Plaintext	Codeword	Plaintext
E CHOS	MISSION	JULIE	TO
A LPHA	LOCATED	KILOS	AT
S IGMA	CONTACT	LIMAS	WITH
T ANGO	RETRN	MILAN	TAKE
BRAVO	SILENCE	NOVEL	BRING
CARLO	COPY	OSCAR	GET
DELTA	START	PAPER	SEND
FOLLY	FINISH	QUICK	MEET
GOOSE	MINUTE	ROMEO	SUPPLY
HOTEL	HOUR	UNITE	LOW
INDIA	TODAY	VIRGO	HIGH

Codeword	Plaintext	Codeword	Plaintext
WHITE	EAST	YALTA	NO
XRAYS	WEST	ZEBRA	NORTH

The keyword selects the first four codewords and plaintext pairs, then the alphabet is continued in order, leaving out the already-used keyword characters. This allows for twenty-six plaintext words. Whenever you wish to encode the message:

P: SEND SUPPLIES TODAY

you could call by radio:

C: LIMAS BRAVO WHITE

Of course, the plaintext word list can be rearranged or changed any way you wish to suit the needs of your mission.

In practical situations, however, you may find that you need more than twenty-six plaintext elements. Suppose we expand Codemaster to two columns.

Codeword	Plaintext	Plaintext
	ECHOS	ALPHA
ECHOS	MISSION	SOUTH
ALPHA	LOCATED	DEGREE
SIGMA	CONTACT	MAN
TANGO	RETRN	SQUAD
BRAVO	SILENCE	PLATOON
CARLO	COPY	COMPANY
DELTA	START	BN
FOLLY	FINISH	BRIGADE
GOOSE	MINUTE	ARMY
HOTEL	HOUR	FIGHTER
INDIA	TODAY	BOMBER
JULIE	TO	YES

Codemaster System

Codeword	Plaintext	Plaintext
KILOS	AT	GREEN
LIMAS	WITH	RED
MILAN	TAKE	BLUE
NOVEL	BRING	DAY
OSCAR	GET	PRIORITY
PAPER	SEND	NIGHT
QUICK	MEET	ORANGE
ROMEO	SUPPLY	VEHICLE
UNITE	LOW	FUEL
VIRGO	HIGH	AIRCRFT
WHITE	EAST	ARMOR
XRAYS	WEST	AB
YALTA	NO	MEDICAL
ZEBRA	NORTH	INFNTRY

Notice the codewords *ECHOS* above the first column and *ALPHA* above the second column. The plaintext *SILENCE* would now be identified by the coordinates *BRAVO ECHOS*. *BN* would be identified by *DELTA ALPHA*.

The message "Send supplies today" would now be sent:

C: PAPER ECHOS ROMEO ECHOS INDIA ECHOS

We can add additional columns—up to twenty-six—for a total of 676 plaintext words.

The manual arrangement used above would produce a reliable codeword system that would be secure for a considerable time and could be changed easily, but a superior system would use a keyword that is seemingly unrelated to the order of the coordinates. A computer is required for this in order to generate a reliable rearrangement of the codewords.

The program below, written in BASIC for Casio, Sharp, and Radio Shack pocket computers, will construct a virtually unbreakable codeword system. It can produce codeword/plaintext arrays in any dimensions from 1×1 to 26×26. The array dimensions are controlled by the variables *I* and *H* in line ten. As the program is written, it is set for an array of 26×2 or a total of fifty-two plaintext words. In this case, there are twenty-six rows and two columns. If you use larger arrays, be sure you have adequate plaintext words typed in the data listing located from line 800 to 929. If there is not enough data for the array you set, you will get an error that will stop execution of the program. Any plaintext words can be changed by typing over those in memory or reentering them.

A keyword should consist of alphabetical characters only and be six characters in length. An example is *QWERTY*. *THOMAS*, *BETTIE*, *SYSTEM*, or *LETTER* would also work fine. Entries such as *&HB1M4* may work, but they are unreliable. Stick with any group of six alphabetical characters, including words with repetitive letters.

Remember that the commands and syntax of various versions of BASIC differ slightly so you may get errors when you enter and attempt to run these programs on certain computers. I am including a similar program in IBM/GW BASIC which should give you a standard reference. The IBM/GW BASIC should run on any IBM compatible.

If you enter the program on a Casio, Radio Shack, or similar pocket micromicro, use the keyword *QWERTY* to test it. If you leave the variables in line ten as written, you will have an array, or table, of twenty-six rows and two columns. The variable is *H*=26 and *I*=2. The table below is the relationships you should get with *QWERTY* as the keyword. If you have to debug your program, always use this or a keyword with which you have experience so that the results can be compared with as many known factors as possible. Be sure to get the program working perfectly before you attempt to change anything. If you use any other BASIC, you may get a variation of the pattern with the keyword *QWERTY*. Also, a rigid check should be made before attempting to use two or more different computer models or brands in a cipher system for your unit. Many times different models will operate perfectly together, but check compatibility with many keywords and short messages.

Codewords	TANGO	QUICK
TANGO	MISSION	SOUTH
QUICK	LOCATED	DEGREE
XRAYS	CONTACT	MAN
JULIE	RETRN	SQUAD
YALTA	SILENCE	PLATOON
ECHOS	COPY	COMPANY
FOLLY	START	BN
MILAN	FINISH	BRIGADE
ALPHA	MINUTE	ARMY
INDIA	HOUR	FIGHTER
WHITE	TODAY	BOMBER
VIRGO	TO	YES
HOTEL	AT	GREEN
SIGMA	WITH	RED
NOVEL	TAKE	BLUE
KILOS	BRING	DAY
CARLO	GET	PRIORTY
LIMAS	SEND	NIGHT
ROMEO	MEET	ORANGE
BRAVO	SUPPLY	VEHICLE
OSCAR	LOW	FUEL
ZEBRA	HIGH	AIRCRFT
PAPER	EAST	ARMOR
UNITE	WEST	AB
DELTA	NO	MEDICAL
GOOSE	NORTH	INFNTRY

If you were to change H to 2 and I to 26 in line ten, still using the same keyword, QWERTY, you will get a table like this:

Codewords	TANGO	QUICK	XRAYS	JULIE	YALTA	ECHO	FOLLY
TANGO	MISSION	CONTACT	SILENCE	START	MINUTE	TODAY	AT
QUICK	LOCATED	RETRN	COPY	FINISH	HOUR	TO	WITH
Codewords	MILAN	ALPHA	INDIA	WHITE	VIRGO	HOTEL	SIGMA
TANGO	TAKE	GET	MEET	LOW	EAST	NO	SOUTH
QUICK	BRING	SEND	SUPPLY	HIGH	WEST	NORTH	DEGREE
Codewords	NOVEL	KILOS	CARLO	LIMAS	ROMEO	BRAVO	OSCAR
TANGO	MAN	PLATOON	BN	ARMY	BOMBER	GREEN	BLUE
QUICK	SQUAD	COMPANY	BRIGADE	FIGHTER	YES	RED	DAY
Codewords	ZEBRA	PAPER	UNITE	DELTA	GOOSE		
TANGO	PRIORTY	ORANGE	FUEL	ARMOR	MEDICAL		
QUICK	NIGHT	VEHICLE	AIRCRFT	AB	INFNTRY		

There are two *rows* and twenty-six *columns*, plus the assignment of the pairs is different.

Still using QWERTY as the keyword, if you set H to 10 and I to 10 in line ten, you will get an array like this:

	TANGO	QUICK	XRAYS	JULIE	YALTA	ECHOS	FOLLY	MILAN	ALPHA	INDIA
TANGO	MISSION	TODAY	LOW	PLATOON	BLUE	MEDICAL	THREE	B	L	V
QUICK	LOCATED	TO	HIGH	COMPANY	DAY	INFNTRY	FOUR	C	M	W
XRAYS	CONTACT	AT	EAST	BN	PRIORTY	TRUCK	FIVE	D	N	X
JULIE	RETRN	WITH	WEST	BRIGADE	NIGHT	JEEP	SIX	E	O	Y
YALTA	SILENCE	TAKE	NO	ARMY	ORANGE	FOOD	SEVEN	F	P	Z
ECHOS	COPY	BRING	NORTH	FIGHTER	VEHICLE	AMMO	EIGHT	G	Q	THIS
FOLLY	BEGIN	GET	SOUTH	BOMBER	FUEL	NEED	NINE	H	R	THAT
MILAN	END	SEND	DEGREE	YES	AIRCRAFT	ARTLRY	TEN	I	S	THE
ALPHA	MINUTE	MEET	MAN	GREEN	ARMOR	ONE	ZERO	J	T	THEM
INDIA	HOUR	SUPPLY	SQUAD	RED	AB	TWO	A	K	U	?

Be sure that you do not use any of the BASIC *reserved words* as plaintext or keywords. Notice that I have written *RETRN* for *RETURN* in one of the tables above. *RETURN* is a reserved word in BASIC and can cause an error if you use it in the wrong place.

Here is the BASIC program for Casio, Radio Shack, Sharp, and other pocket micromicros:

```
10  CLEAR : DIMA$(26,26),B$(26),U$(26):H = 26 : I = 2
15  PRINT "WAIT";
20  FOR N = 1 TO I
22  FOR X = 1 TO H
25  READ A$ (N,X)
27  NEXT X : NEXT N
33  PRINT
35  PRINT "PRESS EXE FOR MENU", : Y$ = ""
37  INPUT "NEW KEYWORD Y/N", Y$ : IF Y$ = "Y"
    THEN 700
50  INPUT "ENCODE Y/N", Y$ : IF Y$ = "Y" THEN 80
55  INPUT "DECODE Y/N", Y$ : IF Y$ = "Y" THEN 180
57  IF Y$ <> "Y" THEN 35: REM "IS NOT" IS
    REPRESENTED IN VARIOUS WAYS
80  IF G = 0 THEN PRINT "NO KEY! PRESS EXE", :
    GOTO 35
90  INPUT "ENTER PLAINTEXT", P$ : PRINT CSR0;
    P$;
101 IF P$ = "" THEN 35
105 FOR N = 1 TO I : FOR X = 1 TO H : F = 0
106 IF A$(N,X) = P$ THEN F = 1 : PRINT CSR10; B$(X);
    CSR17; B$(N)
107 IF F = 1 THEN 80
110 NEXT X : NEXT N
115 FOR X = 1 TO 6 : BEEP : NEXT X
120 IF F = 0 THEN PRINT "ERROR*PRESS EXE", :
    GOTO 80
180 IF G = 0 THEN 80
```

```
200 INPUT "ENTER CODEWORD 1", K$: IF K$ = ""
    THEN 35
208 INPUT "ENTER CODEWORD 2", J$: IF J$ = ""
    THEN 35
209 F = 0
212 FOR N = 1 TO 26
216 IF J$ = B$(N) THEN X = N: F = F + 1
220 IF K$ = B$(N) THEN L = N: F = F + 1
224 NEXT N
226 IF A$(X,L) = ""; PRINT "ERROR*PRESS EXE",:
    GOTO 200 : REM THE ";" SYMBOL IS SOMETIMES
    USED FOR "THEN"
230 IF F = 2; PRINT CSR0; K$; CSR6; J$; CSR16; A$(X,L):
    GOTO 200
234 PRINT "ERROR*PRESS EXE",: GOTO 200
700 RESTORE 940 : C$ = "3": FOR N = 1 TO 26
702 READ U$(N): NEXT N
704 INPUT "ENTER KEYWORD",$: IF $ = "" THEN 35
    : REM $ IS A SPECIAL CHARACTER IN CASIO
    BASIC
705 PRINT "WAIT"; : G = 0
706 FOR X = 1 TO 6: W$ = MID$(X,1)
708 FOR N = 1 TO 26: F = 0
710 IF W$ = U$(N) THEN Z = N + X: F = 1
711 IF F = 1; IF Z ⩾10; Z = Z − 10: GOTO 711
714 IF F = 1; IF Z = 0; C$ = C$ + STR$(Z): GOTO 716
715 IF F = 1 THEN C$ = STR$(Z) + C$
716 NEXT N : NEXT X
718 V = VAL(C$)
721 Z = V*.123456789: O = 7: DIM E(26)
724 FOR X = 1 TO 26
728 IF O = 500 THEN O = 1.57697
734 O = O + 1 : S = (997*Z + (O↑2))/199
736 Z = FRAC(S): Z = INT(Z*100)
738 IF Z>26 THEN 728
740 IF Z<1 THEN 728
```

```
742 E(X) = Z
744 T = X - 1
746 FOR R = 0 TO T
748 IF Z = E(R) THEN 728
750 NEXT R
752 NEXT X
754 RESTORE 930
758 FOR N = 1 TO 26: READ B$(E(N)): NEXT N
759 G = 1: FOR N = 1 TO 6: BEEP: NEXT N
760 PRINT: PRINT"KEY INSTALLED*PRESS EXE";:
    GOTO 35
800 DATA MISSION,LOCATED,CONTACT,RETRN,
    SILENCE,COPY
801 DATA START,FINISH,MINUTE,HOUR,TODAY,TO
802 DATA AT,WITH,TAKE,BRING,GET,SEND
803 DATA MEET,SUPPLY,LOW,HIGH,EAST,WEST
804 DATA NO,NORTH,SOUTH,DEGREE,MAN,SQUAD
805 DATA PLATOON,COMPANY,BN,BRIGADE,ARMY,
    FIGHTER
806 DATA BOMBER,YES,GREEN,RED,BLUE,DAY
807 DATA PRIORTY,NIGHT,ORANGE,VEHICLE,FUEL,
    AIRCRFT
808 DATA ARMOR,AB,MEDICAL,INFNTRY,TRUCK,JEEP
809 DATA FOOD,AMMO,NEED,ARTLRY,ONE,TWO
810 DATA THREE,FOUR,FIVE,SIX,SEVEN,EIGHT
811 DATA NINE,TEN,ZERO,A,B,C
812 DATA D,E,F,G,H,I
813 DATA J,K,L,M,N,O
814 DATA P,Q,R,S,T,U
815 DATA V,W,X,Y,Z,THIS
816 DATA THAT,THE,THEM,?
930 DATA ALPHA,BRAVO,CARLO,DELTA,ECHOS,FOLLY
931 DATA GOOSE,HOTEL,INDIA,JULIE,KILOS,LIMAS
932 DATA MILAN,NOVEL,OSCAR,PAPER,QUICK,
    ROMEO
933 DATA SIGMA,TANGO,UNITE,VIRGO,WHITE,XRAYS
```

```
934 DATA YALTA,ZEBRA
940 DATA A,B,C,D,E,F,G,H,I,J,K,L,M
941 DATA N,O,P,Q,R,S,T,U,V,W,X,Y,Z
```

Here is the same program listed for IBM/GW BASIC.

```
10 CLEAR : DIM A$(26,26),B$(26),U$(26):H = 26 : I = 2
20 FOR N = 1 TO I
22 FOR X = 1 TO H
25 READ A$ (N,X)
27 NEXT X : NEXT N : REM BOOKCODE 12.30
33 PRINT : REM COPYRIGHT 1977, 1986, 1989
   H NICKELS, VER 12.30
35 CLS : PRINT " * MENU FOLLOWS * ", : PRINT :
   Y$ = ""
37 INPUT "NEW KEYWORD Y/N ", Y$ : PRINT :
   IF Y$ = "Y" THEN CLS : GOTO 700
50 PRINT: INPUT "ENCODE Y/N ", Y$ : PRINT :
   IF Y$ = "Y" THEN CLS : GOTO 80
55 PRINT: INPUT "DECODE Y/N ", Y$ : PRINT :
   IF Y$ = "Y" THEN CLS : GOTO 180
56 PRINT: INPUT "QUIT Y/N ", Y$ : PRINT :
   IF Y$ = "Y" THEN CLS: END
57 CLS: IF Y$ <> "Y" THEN 35 : ' <> "is not equal
   to" is represented in various ways
80 PRINT: IF G = 0 THEN PRINT : INPUT "NO KEY!
   PRESS ENTER ",Y$
82 IF G = 0 THEN 35
84 PRINT
90 INPUT "ENTER PLAINTEXT ", P$ : PRINT
101 IF P$ = "" THEN 35
105 FOR N = 1 TO I : FOR X = 1 TO H : F = 0
106 IF A$(N,X) = P$ THEN F = 1 : PRINT P$, B$(X),
    B$(N) : PRINT
107 IF F = 1 THEN PRINT: GOTO 80
110 NEXT X : NEXT N
```

```
115 FOR X = 1 TO 6 : BEEP : NEXT X
120 IF F = 0 THEN PRINT "*ERROR*", : PRINT :
    GOTO 80
180 IF G = 0 THEN 80
200 INPUT "ENTER CODEWORD 1 ", K$: IF K$ = ""
    THEN 35
208 INPUT "ENTER CODEWORD 2 ", J$: IF J$ = ""
    THEN 35
209 F = 0 : PRINT
212 FOR N = 1 TO 26
216 IF J$ = B$(N) THEN X = N: F = F+1
220 IF K$ = B$(N) THEN L = N: F = F+1
224 NEXT N
226 IF A$(X,L) = "" THEN PRINT: INPUT
    "ERROR*PRESS ENTER ",Y$ : GOTO 200
230 IF F = 2 THEN PRINT K$, J$, A$(X,L) : PRINT :
    GOTO 200
234 PRINT "**ERROR**" : PRINT : GOTO 200
700 RESTORE 940 : C$ = "3": FOR N = 1 TO 26
702 READ U$(N): NEXT N
704 INPUT "ENTER KEYWORD ", AKEY$ : IF AKEY$
    = "" THEN 35
705 PRINT "WAIT ", : G = 0
706 FOR X = 1 TO 6: W$ = MID$(AKEY$,X,1)
708 FOR N = 1 TO 26: F = 0
710 IF W$ = U$(N) THEN Z = N + X: F = 1
711 IF F = 1 THEN IF Z >= 10 THEN Z = Z - 10:
    GOTO 711
712 TRY$ = STR$(Z) : TRY$ = MID$(TRY$,2,1)
714 IF F = 1 THEN IF Z = 0 THEN C$ = C$ + TRY$
    : GOTO 716
715 IF F = 1 THEN C$ = TRY$ + C$
716 NEXT N : NEXT X
718 V = VAL(C$)
721 Z = V * .123456789: O = 7: DIM E(26)
724 FOR X = 1 TO 26
728 IF O = 500 THEN O = 1.57697
```

Codemaster System

```
734 O = O + 1 : S = (997*Z+(O^2))/199
736 Z = S - INT(S): Z = INT(Z*100)
738 IF Z>26 THEN 728
740 IF Z<1 THEN 728
742 E(X) = Z
744 T = X - 1
746 FOR R = 0 TO T
748 IF Z = E(R) THEN 728
750 NEXT R
752 NEXT X
754 RESTORE 930
758 FOR N = 1 TO 26: READ B$(E(N)): NEXT N
759 G = 1: FOR N = 1 TO 6: BEEP: NEXT N
760 PRINT: INPUT "KEY INSTALLED*PRESS ENTER
    ", Y$ : PRINT : GOTO 50
800 DATA MISSION,LOCATED,CONTACT,RETRN,
    SILENCE,COPY
801 DATA START,FINISH,MINUTE,HOUR,TODAY,TO
802 DATA AT,WITH,TAKE,BRING,GET,SEND
803 DATA MEET,SUPPLY,LOW,HIGH,EAST,WEST
804 DATA NO,NORTH,SOUTH,DEGREE,MAN,SQUAD
805 DATA PLATOON,COMPANY,BN,BRIGADE,ARMY,
    FIGHTER
806 DATA BOMBER,YES,GREEN,RED,BLUE,DAY
807 DATA PRIORTY,NIGHT,ORANGE,VEHICLE,FUEL,
    AIRCRFT
808 DATA ARMOR,AB,MEDICAL,INFNTRY,TRUCK,JEEP
809 DATA FOOD,AMMO,NEED,ARTLRY,ONE,TWO
810 DATA THREE,FOUR,FIVE,SIX,SEVEN,EIGHT
811 DATA NINE,TEN,ZERO,A,B,C
812 DATA D,E,F,G,H,I
813 DATA J,K,L,M,N,O
814 DATA P,Q,R,S,T,U
815 DATA V,W,X,Y,Z,THIS
816 DATA THAT,THE,THEM,?
930 DATA ALPHA,BRAVO,CARLO,DELTA,ECHOS,FOLLY
931 DATA GOOSE,HOTEL,INDIA,JULIE,KILOS,LIMAS
```

```
932 DATA  MILAN,NOVEL,OSCAR,PAPER,QUICK,
    ROMEO
933 DATA  SIGMA,TANGO,UNITE,VIRGO,WHITE,XRAYS
934 DATA  YALTA,ZEBRA
940 DATA  A,B,C,D,E,F,G,H,I,J,K,L,M
941 DATA  N,O,P,Q,R,S,T,U,V,W,X,Y,Z
```

The line numbers in the IBM/GW BASIC version are roughly analogous to the micromicro version, enabling you to compare the various differences in these BASICs. Amiga, Commodore, Apple, and other BASICs have other differences.

CHAPTER 8

Field Message Systems

■

Almost invariably, a new student of cryptology will dream of constructing a revolutionary, unbreakable cipher system. If the intended use of the system is of a serious nature, this is an extremely dangerous undertaking. This is true because probably every possible system or combination of systems has been used, proposed, or pondered by expert cryptographers who have spent their lives twisting and turning the basic schemes into a "new" system.

Even the most modern systems, such as the Data Encryption Standard (DES), is nothing more than a complex transposition and substitution product cipher with a long key. The United States government commissioned IBM to produce the system for use in the transmission of general data within the government and by large private interests. The system has been controversial since it was learned that there may be secret "tricks" known only to the insiders that would allow easy decipherment of material regardless of the complexity of the so-called "ultimate" system. This clouded system can be purchased today in hardware or software form and is being used by a growing number of unwary organizations.

If all systems have been invented by earlier cryptologists, then rather than attempt to construct a new, unbreakable system, the serious user should make every effort to understand the system he is about to use. Even though the system he has is a common one, he may then be able to alter the internal keys or part of the algorithm so that it would be secure for his use. A general rule of the professional cryptologist is this: the adversary knows the generic system in use, but what he does not know is the key, the method of applying the key, and the internal keys. Nor does he know when and how the keys will be changed. This is then the strong hand held by the cryptographer. Obviously, he should concentrate on these factors rather than attempt to construct an entirely new system.

These ideas rule out using any system that is difficult to change—especially for field use by a relatively small organization with rudimentary equipment and whose primary mission is not cryptography. The selected systems must allow changes to be made quickly and uniformly even in difficult circumstances.

MANUAL SYSTEMS

The first scheme, refined by the author during the Korean Police Action, was designed in an emergency to temporarily replace, on the battalion and company level, some of the ancient and rusted World War II M-209 rotor code machines that the U.S. Army was trying to use at the time. It has a simple key arrangement, can be learned in a few minutes, and requires little more than a pencil and paper. At the time, we called it the *Pandora's Box* system or just *Pandora*.

First, a block of twenty-five squares is drawn out and the keyword is written in using left to right, then down. The keyword is REDHOT.

R E D H O
T

Then write out the alphabet.

A B C D E F G H I J K L M N O P Q R S T U V W X Y Z

Now, erase or cross out any characters found in the keyword·
A B C D̶ E̶ F G H̶ I J K L M N O̶ P Q R̶ S T̶ U V W X Y Z
Copy the remaining characters of the alphabet into the box.

R E D H O
T A B C F
G I J K L
M N P Q S
U V W X *

*Y/Z go in
this position.

Now, copy the first column to a sixth column.

R E D H O R
T A B C F T
G I J K L G
M N P Q S M
U V W X * U

*Y/Z go in
this position.

Move a copy of the bottom row to the top.

```
U V W X * U
R E D H O R
T A B C F T
G I J K L G
M N P Q S M
U V W X * U
```

*Y/Z go in
this position.

This is the plaintext we must encipher:

P: attack hill six zero
P: attac khill sixze roaei
(Plaintext with nulls added to make groups of five evenly divisible by two.)

We will use two characters to replace each plaintext character. Think of the *a* as forming the lower left corner of a box of four letters. We can replace the *a* with *ED*, *DB*, *BD*, *DE*, *EB*, or *BE*. To decipher, we would look for the four-character box that has any of these combinations, and it would invariably point to the *a* in the lower left corner. Even the combination *AD* or *DA* could be used as ciphercharacters.

```
P: a t t a c  k h i l l  s i x z e  r o a e i
C: E R A D O  C X J T G  M B Q U V  E Y B W B
C: B A R B F  F O B F T  L A Z F D  U U E D I
```

Notice that we place the ciphercharacter *EB* under the *a*. This allows, as the message is transmitted, for the *EB* to be split, effectively fractionating the *a*, putting part of the *a* in one place in the message and the balance in another. This is also a form of transposition. The first five-letter group sent out, then, is *ERADO*.

Field Message Systems

Look at the way the second *a* is enciphered. We have taken care to use a different combination for the cipher characters. The third *a* has an even different cipher pair.

The allowable lengths of messages must be agreed upon, of course, in advance so that the ciphercharacter pairs can be reassembled correctly. In our case, we stipulated that the character number was to be evenly divisible by two so that the ciphertext could be split in half and easily realigned. Other specifications could just as easily be assigned. In fact, all methods should be made different and, if possible, a bit more complex than we are using in our examples.

Again, to decipher, the ciphertext must be restacked in the agreed-upon scheme, then each pair taken off.

Effectively, this is a *product* cipher. That is, it is a combination of transposition and multiple substitution.

Many variations can be used with this basic system. Here is a list of possible alternatives:

1. Instead of using the lower left corner for the ciphercharacter, use the upper right, or the upper left, or the lower right.

2. Insert the keyword in another position, then follow with the balance of the alphabet in reverse order.

3. Keywords need not be restricted to words that have no repetitive letters. The repetitive letters can be omitted. For example use GEORGEWASHINGTON as a keyword. GEORWASHINT would result. The longer the codeword the better. You must select phrases that can be easily remembered and spelled by your associates.

4. The initial box could be formed using the length of the keyword to govern the number of columns. Keyword length between six and thirteen would be best. This one thing would improve security considerably. Then the 5×5 box could be constructed by taking off the columns and forming rows. The initial box would look like this:

```
R E D H O T
A B C F G I
J K L M N P
Q S U V W X
    *
```

*Y/Z go in
this position.

When transposed it would look like this:

```
R A J Q *
E B K S D
C L U H F
M V O G N
WT I P X
```

*Y/Z go in
this position.

Compare it to the original box shown below:

```
R E D H O
T A B C F
G I J K L
MN P Q S
U V W X *
```

*Y/Z go in
this position.

Obviously, the disarrangement of the alphabet is desirable and could add another layer to the difficulties presented to an adversary.

RECOMBINATION OR BLACKBOX SYSTEM

This scheme is similar to the above field cipher, but it does not require that the ciphertext grow to twice the size of the plaintext. This is accomplished by recombining the split fractional parts into a new, single cipher character.

104

Field Message Systems

We will use the keyword *SIXPACK*. The plaintext will be:

P: bluebird in singapore
P: blueb irdin singa porex

Construct the keybox.

```
S I X P A C K
B D E F G H J
L M N O Q R T
U V W *
```

* is Y/Z.

Transpose to a 5×5 box.

```
S B L U I
D M V X E
N W P F O
* A G Q C
H R K J T
```

* is Y/Z.

Add a coordinate index along the top and side.

```
  A B C D E
A S B L U I
B D M V X E
C N W P F O
D * A G Q C
E H R K J T
```

* is Y/Z.

Encipher the plaintext b as its coordinates *AB*, but as in PANDORA'S BOX, place the *AB* below the b.

P: b l u e b i r d i n s i n g a p o r e x
C: A
C: B

Continuing:

P: b l u e b i r d i n s i n g a p o r e x
C: A A A B A A E B A C A A C D D C C E B B
C: B C D E B E B A E A A E A C B C E B E D

Now, take out the pairs from left to right.

AA AB AA EB AC AA CD DC CE BB BC DE BE BA EA
AE AC BC EB ED

Recombine them, using them as coordinates in the same box
that was used to obtain them. *AA* will be S.

AA AB AA EB AC AA CD DC CE BB BC DE BE BA EA
 S   B   S   R   L   S   F   G   O   M   V   C   E   D   H
AE AC BC EB ED
 I   L   V   R   J

This would be the final ciphertext:

C: SBSRL SFGOM VCEDH ILVRJ

As always, decipherment is the reverse of encipherment.
  This recombination scheme is an excellent field cipher if
you are limited to pencil and paper. It is breakable, but
considerable time and skill would be required. It is suitable
for short messages, especially if nulls are included in the
plaintext and the keyword is changed often.

CRACKER-BOX CIPHER

  The keyword for this manual cipher is a number that is
expanded to form a pseudorandom series. We will use, in

this case, a date: January 17, 1983 or 1171983.
   The plaintext is:

P: send tiger to home base

First, determine the length of the message and add nulls to
make the total length a multiple of five.

P: sendt igert ohome basex

The keyword is then added to itself.

K: 1171983
   1171983
   2343966

Only the answer is used. There are seven digits. We need
twenty to encipher the message. Add the answer to itself.

   2343966
   2343966
   4687932

Now we have fourteen digits in the two answers. In order
to get twenty we must do another cycle.

   4687932
   4687932
   9375864

Put the answers only in a box. Align to the *left* if the number
of digits in any row is not the same as the others.

   2343966
   4687932
   9375864

Now take off the numbers starting with the left column, reading down, then the second column, and so forth.

   2 4 9 3 6 3 4 8 7 3 7 5 9 9 8 6 3 6 6 2 4

Now we must "condition" these numbers so that they will be a series of numbers with values of 1 to 26 inclusive. First, to obtain an even distribution, add 00, 10, and 20 successively to each number in the series like this:

   2   4   9   3   6   3   4   8   7   3   7   5   9   9   8   6   3
  00   10   20   00   10   20   00   10   20   00   10   20   00   10   20   00   10
   2   14   29   3   16   23   4   18   27   3   17   25   9   19   28   6   13
  6   6   2   4
 20   00   10   20
 26   6   12   24

If any of the answers are less than 1 (a zero is possible), add 26. If any of the answers are more than 26, then subtract 26.

   2   4   9   3   6   3   4   8   7   3   7   5   9   9   8   6   3
  00   10   20   00   10   20   00   10   20   00   10   20   00   10   20   00   10
   2   14   29   3   16   23   4   18   27   3   17   25   9   19   28   6   13
            -26   -26   -26
             3   1   2
  6   6   2   4
 20   00   10   20
 26   6   12   24

We now have this series of numbers derived from the keynumber:

   2   14   3   3   16   23   4   18   1   3   17   25   9   19   2   6   13
 26   6   12   24

Next, convert the alphabet to numbers so that the plaintext can be merged with the derived numbers.

a	b	c	d	e	f	g	h	i	j	k	l	m	n	o	p	q	r	s	t	u
1	2	3	4	5	6	7	8	9	10	11	12	13	14	15	16	17	18	19	20	21

v	w	x	y	z
22	23	24	25	26

Write the numerical value of each character below it.

```
s    e   n   d   t    i   g   e   r   t    o    h   o    m    e
19   5  14   4  20    9   7   5  18  20   15    8  15   13    5
b    a   s   e   x
2    1  19   5  24
```

Now write the series of numbers derived from the key-number below those and add them.

```
s    e   n   d   t    i   g   e   r   t    o    h   o    m   e
19   5  14   4  20    9   7   5  18  20   15    8  15   13   5
 2  14   3   3  16   23   4  18   1   3   17   25   9   19   2
21  19  17   7  36   32  11  23  19  23   32   33  24   32   7
b    a   s   e   x
2    1  19   5  24
6   13  26   6  12
8   14  45  11  36
```

Again, using the rule above, subtract 26 from any answer equal to more than 26; add 26 to any number less than 1.

```
s    e   n   d   t    i   g   e   r   t    o    h   o    m   e
19   5  14   4  20    9   7   5  18  20   15    8  15   13   5
 2  14   3   3  16   23   4  18   1   3   17   25   9   19   2
21  19  17   7  36   32  11  23  19  23   32   33  24   32   7
               -26 -26                  -26 -26      -26
               10   6                     6   7        6
```

```
b   a   s   e   x
2   1  19   5  24
6  13  26   6  12
8  14  45  11  36
       -26     -26
        19      10
```

These, then, are the cipher numbers:

21 19 17 7 10   6 11 23 19 23   6 7 24 6 7   8 14 19 11 10

Convert them to cipherletters using the numbered alphabet.

```
21 19 17  7 10   6 11 23 19 23   6  7 24  6  7
 U  S  Q  G  J   F  K  W  S  W   F  G  X  F  G
 8 14 19 11 10
 H  N  S  K  J
```

The cipher ready to be sent out is:

C: USQGJ FKWSW FGXFG HNSKJ

POCKET-CALCULATOR CIPHER

An excellent source of pseudorandom numbers is the pocket calculator. We can use the format of the cracker-box cipher but with an easier method of generating the numbers we need.

Scientific calculators are a gold mine of numbers, but even the lowest priced generic calculators ordinarily have several functions that can produce long strings of real numbers from fairly simple inputs. Mathematically speaking, real numbers are those that are not just whole integers like 1, 4, and 9 but can have fractional values like 1.45, 5.6528910, and 89.93827465. Any function on the calculator that generates long strings of these numbers is a candidate for use in a cipher system.

Field Message Systems

Using any pocket calculator, enter the date that we used in the cracker-box cipher: 1171983. Now press the square root function key. You should get 1082.581637 if your calculator will display 10 digits, 1082.5816 if it is a $4.99 variety. Press the square root key again, effectively taking the square root of the previous root. You should have 32.902608. Press it again. 5.7360795. This gives you enough numbers to encipher the message used in the cracker cipher.

Ignore all decimal points and any other symbols that may appear on the screen of the pocket calculator. Set the numbers up in the box seven across, according to the keynumber.

```
1082581
6329026
0857360
795
```

Now take the numbers off according to the scheme used in cracker.

1 6 0 7 0 3 8 9 8 2 5 5 2 9 7 5 0 3 8 2 6 1 6 0

They can now be processed as in the cracker system.

This series of numbers would be virtually impossible to duplicate without the key. Even if the adversary knows the system you are using, it is unlikely that the cipher could be broken without the key unless he has many copies of matching plaintext and ciphertext.

Numerous other schemes to obtain numbers can be used, but you should test them to make sure that they will yield enough "quality" numbers. As you have probably noticed, the method we have used, if continued, produces smaller and smaller numbers until the answers get closer and closer to one. At this point, numerous zeros appear in each answer. Although these numbers can still be used, it may be best to develop a better method if many numbers will be needed.

One possible system would be to use the following algorithm. Using the keynumber 1171983, first obtain its square root. Now, take the first four digits after the decimal point in the answer, enter them, and take the square root. Using that answer, take the first four digits after the decimal point as before, enter them, and take the square root, and so on and on. Below is the method and run of numbers you could use if you needed sixty numbers for your message:

Starting with 1171983, the square root would be: 1082.5816

Now enter 5816 (do not use the decimal point) and press the square root function key. You get 76.262704. Put it below your last answer. Take the 2627 out of it and continue as before.

1082.5816
76.262704
51.254268
50.418251
64.668385
81.749618
86.579443
76.118329

Remove the decimal points; align to the left.

1082581
6762627
0451254
2685041
8251646
6838581
7496188
6579443
7611832
9

Now the number series 1 6 0 2 8 6 7 6 7 9 0 7 4 6. . .can be obtained, conditioned, and used to merge the plaintext into ciphertext. This produces a top-notch manual system, virtually unbreakable.

Other function keys and schemes can be used. In fact, the author has devised more than two hundred systems based on the functions available on a generic scientific calculator. Extremely secure systems can be devised using several functions mixed in a pattern. For example, you could first take the square root of the key, then extract several of the numbers following the decimal point, then square that number. Then, extract the numbers after the decimal point in that answer, then take the square root of that number—and so on. If you use a scientific calculator, you have many more functions that could be alternated or mixed into the scheme.

Keywords can be used as well as key numbers. Using one scheme, the letters are simply changed to number values. GEORGE would equal 7 5 15 18 7 5 or 75151875.

THE DIABLO (VERSION 6.26C)

The following program can be loaded into some of the smallest pocket micromicros, including the Casio FX-795P, the Tandy PC-6, all Sharp models, and others. It uses a ten-character key, combines transposition and polyalphabetical substitution into a very complex product cipher, and will accept a message of unlimited length. Further, it is easily modified for the addition of an internal key that would make it useful in an organization that has several levels of security. For example, it could be operated in a system as shown below, yet each member of the group would use the same external key during a given period.

```
┌─────────────────────────────────────────────┐
│           Captain A holds the                 │
│    masterkey—he can encipher and              │
│         decipher all messages                 │
└─────────────────────────────────────────────┘
        │                  │              │
┌──────────────┐  ┌──────────────┐  ┌──────────────┐
│  Lt. B can   │  │  Lt. C can   │  │  Lt. D can   │
│   encipher   │  │   encipher   │  │   encipher   │
│     and      │  │     and      │  │     and      │
│   decipher   │  │   decipher   │  │   decipher   │
│  only with   │  │  only with   │  │  only with   │
│  Captain A   │  │  Captain A   │  │  Captain A   │
└──────────────┘  └──────────────┘  └──────────────┘
```

This is the program listing for the pocket micromicros:

```
5  REM       PROGRAM DIABLO VERSION 6.26C
10 CLEAR : DIM A$(5),B$(5),E(5),U$(27),K(10),F(5),
   J(5),X$(5)
11 RESTORE 940 : FOR N = 1 TO 26
12 READ U$(N): NEXT N
15 REM COPYRIGHT 1983 1989 H. NICKELS
30 REM       MENU BLOCK
33 PRINT
35 PRINT " * PRESS EXE FOR MENU * ", : PRINT :
   Y$ = ""
36 PRINT
37 INPUT "NEW KEYWORD Y/N ", Y$: IF Y$ = "Y"
   THEN GOTO 904
40 PRINT
```

```
50 INPUT "ENCODE Y/N ", Y$ : IF Y$ = "Y" THEN
   GOTO 70
52 PRINT
55 INPUT "DECODE Y/N ", Y$ : IF Y$ = "Y" THEN
   GOTO 800
56 PRINT
57 INPUT "QUIT Y/N ", Y$ : IF Y$ = "Y" THEN END
58 IF Y$ <> "Y" THEN 35 : REM "IS NOT EQUAL
   TO" IS STATED SEVERAL WAYS
60 PRINT
70 REM      ENCIPHER BLOCK
80 IF G = 0 THEN PRINT : INPUT "NO KEY!
   PRESS EXE ",Y$
82 IF G = 0 THEN 35
84 PRINT
90 INPUT "ENTER PLAINWORD ", $
101 IF $ = "" THEN 90
110 IF $ = "XXX" THEN 10
120 IF LEN($)<> 5 THEN 90
125 PRINT "P: ";$; " ";
130 FOR N = 1 TO 5
140 A$(N) = MID$(N,1)
150 NEXT N
155 REM      ENCIPHER TRANSPOSE BLOCK
724 FOR X = 1 TO 5
728 IF INT(O) = 30000 THEN O = 1.57697
734 O = O + 1 : Q = O↑1.123 : S = (997*Z+Q + K(X))
    * .0050251256
735 REM      THE POWER OPERATOR IS STATED
             SEVERAL WAYS
736 Z = S – INT(S): Z = INT(Z*100): Y=Z
738 IF Y>5 THEN 728
740 IF Y<1 THEN 728
742 E(X) = Y
744 T = X – 1
746 FOR R = 0 TO T
```

```
748 IF Y = E(R) THEN 728
750 NEXT R
752 NEXT X
758 FOR N = 1 TO 5: B$(E(N)) = A$(N): NEXT N
760 FOR X = 1 TO 5
762 REM        ENCIPHER SUBSTITUTION BLOCK
763 O = O + 1: Q=O↑1.123 : S = (997*Z+Q+ K(X+5) )
    *.0050251256
764 Z = S – INT(S): Z=INT(Z*100): Y=Z
765 IF Y>26 THEN 762
766 IF Y<1 THEN 762
768 E(X) = Y: NEXT X
770 FOR N = 1 TO 26
772 FOR R = 1 TO 5
773 IF U$(N) = B$(R) THEN F(R) = N
774 NEXT R : NEXT N
776 PRINT "C: ";
778 FOR N = 1 TO 5
780 H = E(N) + F(N)
782 IF H>26 THEN H = H – 26
786 PRINT U$(H);
788 NEXT N : PRINT"",: GOTO 80
800 REM        DECIPHER BLOCK
802 IF G = 0 THEN PRINT : INPUT "NO KEY!
    PRESS EXE ",Y$
803 IF G = 0 THEN 35
804 REM
813 INPUT "ENTER CIPHERWORD ", $
815 IF $ = "" THEN 813
817 IF $ = "XXX" THEN 10
819 IF LEN($)<> 5 THEN GOTO 813
820 FOR N = 1 TO 5
821 A$(N) = MID$(N,1)
822 NEXT N
823 PRINT "C: ";$;" ";
824 FOR X = 1 TO 5
828 IF INT(O) = 30000 THEN O = 1.57697
```

```
834 O = O + 1 : Q = O↑1.123 : S = (997*Z+Q + K(X))
    * .0050251256
836 Z = S-INT(S): Z = INT(Z*100): Y=Z
838 IF Y>5 THEN 828
840 IF Y<1 THEN 828
842 J(X) = Y
844 T = X - 1
846 FOR R = 0 TO T
848 IF Y = J(R) THEN 828
850 NEXT R
852 NEXT X
860 FOR X = 1 TO 5
862 REM     DECIPHER SUBSTITUTION BLOCK
863 O = O + 1: Q=O↑1.123 : S = (997*Z+Q+ K(X+5) )
    * .0050251256
864 Z = S-INT(S): Z=INT(Z*100): Y=Z
865 IF Y>26 THEN 862
866 IF Y<1 THEN 862
868 E(X) = Y
869 NEXT X
870 REM
872 FOR N = 1 TO 26
874 FOR R = 1 TO 5
875 IF U$(N) = A$(R) THEN F(R) = N
878 NEXT R : NEXT N
880 PRINT "P: ";
882 FOR N = 1 TO 5
884 H = F(N) - E(N)
886 IF H < 1 THEN H = H + 26
890 X$(N) = U$(H)
892 NEXT N
893 REM     DECIPHER TRANSPOSE BLOCK
894 FOR N = 1 TO 5: B$(N) = X$(J(N)): NEXT N
896 FOR N = 1 TO 5: PRINT B$(N);
898 NEXT N : PRINT "",
900 GOTO 800
902 REM     KEY BLOCK
```

```
904 INPUT "ENTER KEYWORD ", $ : IF $ = ""
    THEN 35
905 G = 0
906 FOR X = 1 TO 10 : W$ = MID$(X,1)
908 FOR N = 1 TO 26: F = 0
910 IF W$ = U$(N) THEN Z = N + X: F = 1
911 IF F = 1 THEN IF Z >= 10 THEN Z = Z - 10:
    GOTO 911
912 K(X) = Z
916 NEXT N : NEXT X
921 O = 0
923 FOR X = 1 TO 10 : Z = Z + K(X) : NEXT X
927 G = 1: BEEP
930 PRINT: INPUT "KEY INSTALLED*PRESS EXE ",
    Y$ : GOTO 50
935 REM     DATA BLOCK
940 DATA A,B,C,D,E,F,G,H,I,J,K,L,M
941 DATA N,O,P,Q,R,S,T,U,V,W,X,Y,Z,X,X
```

Following is the listing for DIABLO written for PC
compatibles:

```
 5 REM     PROGRAM DIABLO VERSION 6.26PC
10 CLEAR : DIM A$(5),B$(5),E(5),U$(26),K(10),F(5)
11 FOR N = 1 TO 26
12 READ U$(N): NEXT N
15 REM     COPYRIGHT 1983 1989 H NICKELS
30 REM     MENU BLOCK
33 PRINT
35 CLS : PRINT " * MENU FOLLOWS * ", : PRINT :
   Y$ = ""
36 PRINT
37 INPUT "NEW KEYWORD Y/N ", Y$: PRINT: IF
   Y$ = "Y" THEN CLS: GOTO 904
40 PRINT
50 INPUT "ENCODE Y/N ", Y$ : PRINT : IF Y$ = "Y"
   THEN CLS : GOTO 70
```

```
52 PRINT
55 INPUT "DECODE Y/N ", Y$ : PRINT : IF Y$ = "Y"
   THEN CLS : GOTO 800
56 PRINT
57 INPUT "QUIT Y/N ", Y$ : PRINT : IF Y$ = "Y"
   THEN CLS : END
58 CLS: IF Y$<> "Y" THEN 35
60 PRINT
70 REM     ENCIPHER BLOCK
80 IF G = 0 THEN PRINT : INPUT "NO KEY!
   PRESS ENTER ",Y$
82 IF G = 0 THEN 35
84 PRINT
90 INPUT "ENTER PLAINWORD ", P$ : PRINT
101 IF P$ = "" THEN 90
110 IF P$ = "XXX" THEN 10
120 IF LEN(P$)<> 5 THEN GOTO 90
130 FOR N = 1 TO 5
140 A$(N) = MID$(P$,N,1)
150 NEXT N
155 REM     ENCIPHER TRANSPOSE BLOCK
724 FOR X = 1 TO 5
728 IF INT(O) = 30000 THEN O = 1.57697
734 O = O + 1 : Q = O^1.123 : S = (997*Z+Q = K(X))
    * .0050251256
736 Z = S - INT(S): Z = INT(Z*100): Y = Z
738 IF Y>5 THEN 728
740 IF Y<1 THEN 728
742 E(X) = Y
744 T = X - 1
746 FOR R = 0 TO T
748 IF Y = E(R) THEN 728
750 NEXT R
752 NEXT X
758 FOR N = 1 TO 5: B$(E(N)) = A$(N): NEXT N
760 FOR X = 1 TO 5
762 REM     ENCIPHER SUBSTITUTION BLOCK
```

```
763 O = O + 1: Q = O^1.123 : S = (997*Z + Q + K(X + 5))
    * .0050251256
764 Z = S - INT(S): Z = INT(Z*100): Y = Z
765 IF Y>26 THEN 762
766 IF Y<1 THEN 762
768 E(X) = Y: NEXT X
770 FOR N = 1 TO 26
772 FOR R = 1 TO 5
773 IF U$(N) = B$(R) THEN F(R) = N
774 NEXT R : NEXT N
776 PRINT "CIPHERWORD ";
778 FOR N = 1 TO 5
780 H = E(N) + F(N)
782 IF H > 26 THEN H = H - 26
786 PRINT U$(H);
788 NEXT N : PRINT : GOTO 80
800 REM     DECIPHER BLOCK
802 IF G = 0 THEN PRINT : INPUT "NO KEY!
    PRESS ENTER ",Y$
803 IF G = 0 THEN 35
804 PRINT
813 INPUT "ENTER CIPHERWORD ", P$
815 IF P$ = "" THEN 813
817 IF P$ = "XXX" THEN 10
819 IF LEN(P$)<> 5 THEN GOTO 813
820 FOR N = 1 TO 5
821 A$(N) = MID$(P$,N,1)
822 NEXT N
824 FOR X = 1 TO 5
828 IF INT(O) = 30000 THEN O = 1.57697
834 O = O + 1 : Q = O^1.123 : S = (997*Z + Q + K(X))
    * .0050251256
836 Z = S - INT(S): Z = INT(Z*100): Y = Z
838 IF Y>5 THEN 828
840 IF Y<1 THEN 828
842 J(X) = Y
```

```
844 T = X - 1
846 FOR R = 0 TO T
848 IF Y = J(R) THEN 828
850 NEXT R
852 NEXT X
860 FOR X = 1 TO 5
862 REM     DECIPHER SUBSTITUTION BLOCK
863 O = O + 1: Q=O^1.123 : S = (997*Z+Q+ K(X+5) )
    * .0050251256
864 Z = S-INT(S): Z=INT(Z*100): Y=Z
865 IF Y>26 THEN 862
866 IF Y<1 THEN 862
868 E(X) = Y
869 NEXT X
870 PRINT
872 FOR N = 1 TO 26
874 FOR R = 1 TO 5
875 IF U$(N) = A$(R) THEN F(R) = N
878 NEXT R : NEXT N
880 PRINT "PLAINWORD ";
882 FOR N = 1 TO 5
884 H = F(N) - E(N)
886 IF H <1 THEN H = H + 26
890 X$(N) = U$(H)
892 NEXT N
893 REM     DECIPHER TRANSPOSE BLOCK
894 FOR N = 1 TO 5: B$(N) = X$(J(N)): NEXT N
896 FOR N = 1 TO 5: PRINT B$(N);
898 NEXT N : PRINT
900 GOTO 800
902 REM    KEY BLOCK
904 INPUT "ENTER KEYWORD ", K$ : IF K$ = ""
    THEN 35
905 G = 0
906 FOR X = 1 TO 10 : W$ = MID$(K$,X,1)
908 FOR N = 1 TO 26: F = 0
```

```
910 IF W$ = U$(N) THEN Z = N + X: F = 1
911 IF F = 1 THEN IF Z >= 10 THEN Z = Z - 10:
    GOTO 911
912 K(X) = Z
916 NEXT N : NEXT X
921 O = 0
923 FOR X = 1 TO 10 : Z = Z + K(X) : NEXT X
927 G = 1: BEEP
930 PRINT: INPUT "KEY INSTALLED*PRESS ENTER
    ", Y$ : PRINT : CLS:GOTO 50
935 REM     DATA BLOCK
940 DATA A,B,C,D,E,F,G,H,I,J,K,L,M
941 DATA N,O,P,Q,R,S,T,U,V,W,X,Y,Z
```

DISCO 4 VERSION 5.30

This program, for PC compatibles, features message storage on disk (these can easily be sent in the mails almost anywhere in the world) and hardcopy production of the enciphered text arranged in five-character groups. It allows the use of the full screen for editing, giving you a total of 1,850 possible characters. It is easy to use and has a number of error-trapping controls written in.

```
 10 CLS : KEY OFF : SCREEN 0 : WIDTH 80
 20 CLEAR, 60000
 22 DEF SEG
 25 DIM KY(1850) : REM COPYRIGHT 1984, 1989
    H. NICKELS
 30 FLAGB = 0 : REM DISCO 4 VERSION 5.30
 60 ON ERROR GOTO 0
 70 LOCATE 2,1
 80 PRINT " MENU
 82 PRINT
 90 PRINT "1 ENTER KEY
100 PRINT "2 CLEAR MEMORY
110 PRINT "3 ENTER TEXT
```

```
120 PRINT "4 ENCIPHER TEXT
130 PRINT "5 DECIPHER TEXT
140 PRINT "6 SAVE CIPHERTEXT TO DISK
150 PRINT "7 LOAD CIPHERTEXT TO MEMORY
152 PRINT "8 QUIT
160 LOCATE 15,1 : INPUT "ENTER CHOICE ",CHOICE
170 IF (CHOICE >8) OR (CHOICE <1) THEN CLS :
    GOTO 60
180 ON CHOICE GOTO 1730,220,370,770,1050,1320,
    1540,1840
220 CLS
230 PRINT "WAIT - CLEARING TEXT MEMORY"
250 FOR X = 60000! TO 61839!
260 POKE X, 32 : POKE X + 2000, 32
270 NEXT
280 BEEP
290 CLS : GOTO 60
370 CLS
371 INPUT"ENTERING PLAINTEXT (P) OR
    CIPHERTEXT (C) ? ",TYPETEXT$
372 PRINT "WAIT - CLEARING MEMORY"
373 FOR X = 60000 TO 61839
374 POKE X,32 : POKE X + 2000,32
375 NEXT
376 CLS
380 PRINT "       INSTRUCTIONS" : PRINT
410 PRINT "ENTER OR CORRECT TEXT USING
    CAPS ONLY, THE UP,"
420 PRINT "DOWN, LEFT, AND RIGHT KEYS.
    CURSOR MUST BE "
422 PRINT "AT THE END OF THE TEXT BEFORE
    ENCIPHERMENT. "
424 PRINT "DO NOT LEAVE SPACES WITHIN THE
    TEXT. MAKE "
426 PRINT "THE CHARACTER COUNT A
    MULTIPLE OF FIVE BY "
```

```
428 PRINT "ADDING NULLS IF NECESSARY. "
450 LOCATE 15,1 : PRINT "PRESS ANY KEY TO
    CONTINUE - " ;
460 A$ = INKEY$ : IF A$ = "" THEN 460 ELSE CLS
    : LOCATE ,,1
540 LOCATE 25,1
542 PRINT "TO EXIT TO MENU - PRESS F10" ; :
    LOCATE 1,1,1
580 PLNTXT$ = INKEY$ : IF PLNTXT$ = "" THEN 580
600 IF CSRLIN > 23 THEN LOCATE 23,POS(0)
680 IF (MID$(PLNTXT$,2,1) = "H") AND (CSRLIN > 1)
    THEN LOCATE CSRLIN - 1,POS(0)
690 IF (MID$(PLNTXT$,2,1) = "K") AND (POS(0) > 1)
    THEN LOCATE CSRLIN,POS(0) - 1
700 IF (MID$(PLNTXT$,2,1) = "M") AND (POS(0) < 80)
    THEN LOCATE CSRLIN,POS(0) + 1
710 IF (MID$(PLNTXT$,2,1) = "P") AND (CSRLIN < 23)
    THEN LOCATE CSRLIN + 1,POS(0)
720 IF (MID$(PLNTXT$,2,1) = "D") THEN SIZE =
    60000 + (CSRLIN - 1) * 80 + POS(0) - 2 : CLS :
    GOSUB 2000 : GOTO 60
724 IF (PLNTXT$ > "Z") OR (PLNTXT$ < "A")
    THEN 580
730 POKE 60000 + (CSRLIN - 1)*80 +
    POS(0) - 1,ASC(PLNTXT$) : PRINT PLNTXT$; :
    GOTO 580
770 CLS
850 CLS : PRINT "WAIT - ENCIPHERING"
860 FOR X = 60000 TO SIZE
880 IF PEEK(X) = 32 THEN GOTO 940
890 RAND = KY(X - 59999!)
900 RANUM = INT(RAND *26)
910 WORKTXT = PEEK(X) - RANUM - 64
920 IF WORKTXT < 1 THEN WORKTXT =
    WORKTXT + 26
930 POKE X + 2000!,WORKTXT + 64
940 NEXT
```

```
 960 CLS
 970 FOR X = 60000 + 2000! TO SIZE+2000
 980 PRINT CHR$(PEEK(X));
 990 NEXT X
 992 LOCATE 25,1
 994 INPUT "PRINT HARDCOPY? Y/N ",A$
 996 IF A$ = "Y" THEN MEM = 2000 : CLS: GOSUB 2050
1000 CLS : LOCATE 25,1 : INPUT "PRESS ENTER TO
     RETURN TO MENU",A$
1010 CLS : GOTO 60
1050 CLS
1052 PRINT "WAIT - DECIPHERING"
1140 FOR X = 60000 TO SIZE : RANDA = KY(X - 59999!)
1160 RANUM = INT(RANDA *26)
1170 ENCITXT = PEEK(X+2000) + RANUM - 64
1180 IF ENCITXT > 26 THEN ENCITXT = ENCITXT
     - 26
1190 POKE X,ENCITXT + 64
1200 NEXT
1220 CLS
1230 FOR X = 60000 TO SIZE
1240 IF (PEEK(X) > 90) OR (PEEK(X) < 65) THEN
     POKE X,32
1250 PRINT CHR$(PEEK(X));
1260 NEXT
1270 LOCATE 25,1
1272 INPUT "PRINT HARDCOPY? Y/N ",A$
1274 IF A$ = "Y" THEN MEM = 0 : CLS : GOSUB 2050
1276 CLS : LOCATE 25,1
1278 INPUT "PRESS ENTER TO RETURN TO
     MENU",A$
1280 CLS : GOTO 60
1320 CLS
1322 ON ERROR GOTO 1485
1370 PRINT "SAVE ENCIPHERED MESSAGE TO DISK"
1400 LOCATE 5,1
1410 INPUT "ENTER FILENAME -
```

```
     DRIVE:XXXXXXXX.XXX ",FILENAME$
1420 LOCATE 10,1
1470 PRINT "SAVING FILE ";FILENAME$
1480 BSAVE FILENAME$,62000!,1840! : GOTO 1490
1485 CLS : BEEP : PRINT"DISK OR DRIVE ERROR "
     : RESUME 1370
1490 CLS : LOCATE 25,1
1492 ON ERROR GOTO 0
1494 PRINT "SAVE OKAY – PRESS ANY KEY TO
     RETURN TO MENU"
1500 A$ = INKEY$ : IF A$ = "" THEN 1500 ELSE CLS
     : GOTO 60
1540 CLS
1542 ON ERROR GOTO 1652
1550 PRINT "LOAD ENCIPHERED MESSAGE FROM
     DISK"
1570 LOCATE 5,1
1580 INPUT "ENTER FILENAME
     DRIVE:XXXXXXXX.XXX ",FILENAME$
1590 LOCATE 10,1
1592 SIZE = 61839
1640 PRINT "LOADING FILE ";FILENAME$
1650 BLOAD FILENAME$,62000! : GOTO 1680
1652 CLS:BEEP:PRINT "DISK OR DRIVE ERROR":
     RESUME 1550
1680 CLS : LOCATE 25,1 : PRINT "LOAD OKAY –
     PRESS ANY KEY TO ";
1682 PRINT "RETURN TO MENU "
1684 ON ERROR GOTO 0
1690 A$ = INKEY$ : IF A$ = "" THEN 1690 ELSE CLS
     : GOTO 60
1730 CLS : FLAGKEY = 1
1740 IF FLAGB = 0 THEN GOTO 1770
1750 PRINT "KEY " CEY, "IS NOW INSTALLED.
     ":INPUT "DO YOU WANT NEW KEY? Y/N ",
     NKEY$
1760 IF NKEY$ = "N" THEN CLS : GOTO 60
```

```
1762 CLS
1770 M = 0 : ZZ=0
1772 INPUT "ENTER KEY (FROM 1 TO 10000) ",CEY
1780 IF (CEY<1) OR (CEY>10000) THEN FLAGB = 0
     :CLS : GOTO 1730
1782 CLS: PRINT "INSTALLING KEY - WAIT
1800 ZZ = CEY : FLAGB=1
1810 FOR KEYS = 1 TO 1840
1820 M = M+1 : FF = M*M:Y=(997*ZZ+FF)
     *.0050251256#:ZZ=(Y-INT(Y))
1830 KY(KEYS) = ZZ : NEXT KEYS : CLS :
     BEEP:GOTO 60
1840 CLS : SYSTEM
2000 IF TYPETEXT$ = "P" THEN RETURN ELSE
2010 PRINT "WAIT"
2012 FOR X = 60000 TO 61839
2014 POKE X + 2000, PEEK(X)
2016 NEXT X : CLS
2020 RETURN
2050 COUNT = 0 : COUNTS = 0
2052 ON ERROR GOTO 2200
2054 WIDTH "LPT1:", 80
2056 FOR X = 60000 + MEM TO SIZE + MEM
2058 LPRINT CHR$(PEEK(X));
2060 COUNT = COUNT + 1
2062 IF COUNT = 5 THEN COUNT = 0 : LPRINT
     CHR$(32);
2064 COUNTS = COUNTS + 1
2068 IF COUNTS = 60 THEN COUNTS = 0 : LPRINT
2070 NEXT X : LPRINT CHR$(12) :GOTO 2210
2200 CLS:BEEP:PRINT"PRINTER ERROR" :RESUME 60
2210 RETURN
```

Only capital letters can be entered when you are in the text editor. It will reject all other characters with the exception of the F10 key, which allows you to exit from the text editor.

The keynumber can be any real or integer number from 1 to 10,000, including decimal fractions such as 12.2468024567894534 or .009876543217658453. This allows quite a large number of keys, making it virtually impossible, or at least impractical, for the codebreaker in the field to "run the numbers" against you even if he has the exact program and a fast computer.

A micromicro version of this program can be used to communicate with a PC, but as we have noted earlier, you must run exhaustive tests to make sure that you have compatibility. For example, you will often get errors in trying to incorporate numbers generated in MS-BASIC with TURBO-BASIC or TRUE BASIC. Numbers developed in PASCAL and C can also lead to erroneous results. Unless you are willing to test and experiment, it is best to use only computers and languages that are of the same variety within a network or system.

A FINAL WARNING REGARDING CODEBREAKERS

Many of the systems presented in this manual can be used in a practical communications organization requiring secrecy, even in extremely serious situations. Are they unbreakable? No. As a cryptographer, this should be ever present in your mind. Remember, your enemy may have a copy of this book. Do not forget it for a moment.

APPENDIX
Common Words and Letters

The following words and character groups are the most commonly used. The order of highest use is from the top left down the first column, then to the next column, and so forth. Codebreakers are looking for these words. You, as a cryptographer, should find ways to avoid using them.

WORDS:

THE	OR	WHEN	ONLY
OF	HER	WHAT	ANY
AND	HAD	YOUR	THEN
TO	AT	MORE	ABOUT
A	FROM	WOULD	THOSE
IN	THIS	THEM	CAN
THAT	MY	SOME	MADE
IS	THEY	THAN	WELL
I	ALL	MAY	OLD
IT	THEIR	UPON	MUST
FOR	AN	ITS	US
AS	SHE	OUT	SAID
WITH	HAS	INTO	TIME
WAS	WERE	OUR	EVEN
HIS	ME	THESE	NEW
HE	BEEN	MAN	COULD
BE	HIM	UP	VERY
NOT	ONE	DO	MUCH
BY	SO	LIKE	OWN
BUT	IF	SHALL	MOST
HAVE	WILL	GREAT	MIGHT
YOU	THERE	NOW	FIRST
WHICH	WHO	SUCH	AFTER
ARE	NO	SHOULD	YET
ON	WE	OTHER	TWO

Common Words and Letters

TRIGRAMS:

THE	HER	HIS	ITH
ING	ATE	RES	TED
AND	VER	ILL	AIN
ION	TER	ARE	EST
ENT	THA	CON	MAN
FOR	ATI	NCE	RED
TIO	HAT	ALL	THI
ERE	ERS	EVE	IVE
REA	INE	ORE	ART
WIT	WHI	BUT	NTE
ONS	OVE	OUT	RAT
ESS	TIN	URE	TUR
AVE	AST	STR	ICA
PER	DER	TIC	ICH
ECT	OUS	AME	NDE
ONE	ROM	COM	PRE
UND	VEN	OUR	ENC
INT	ARD	WER	HAS
ANT	EAR	OME	WHE
HOU	DEN	EEN	WIL
MEN	STI	LAR	ERA
WAS	NOT	LES	LIN
OUN	ORT	SAN	TRA
PRO	THO	STE	
STA	DAY	ANY	

DIGRAMS:

TH	CO	DI	US	EI
IN	DE	SI	MO	AD
ER	RA	CA	OM	SS
RE	RO	UN	AI	IL
AN	LI	UT	PR	OS
HE	RI	NC	WE	UL
AR	IO	WI	AC	EM
EN	LE	HO	EE	NS
TI	ND	TR	ET	OT
TE	MA	BE	SA	GE
AT	SE	CE	NI	IR
ON	AL	WH	RT	AV
HA	IC	LL	NA	CT
OU	FO	FI	OL	TU
IT	IL	NO	EV	DA
ES	NE	TO	IE	AM
ST	LA	PE	MI	CI
OR	TA	AS	NG	SU
NT	EL	WA	PL	BL
HI	ME	UR	IV	OF
EA	EC	LO	PO	BU
VE	IS	PA	CH	

SINGLE LETTERS:

E	S	C	W	K
T	R	U	Y	Q
A	H	P	B	X
O	L	F	G	J
N	D	M	V	Z
I				

OTHER NOTES:

The characters E, T, A, O, N, I, S, R, and H make up more than 70 percent of the average text.

The characters T, O, A, W, B, C, D, S, H, F, M, R, I, and Y

form the initial letter of most words.

E, T, D, S, N, R, and Y form the final letter of most words.

A special note about q. With a few exceptions—mostly abbreviations such as qrp and other unusual uses, q is invariably followed by a u—so it is obvious that the use of a q or u will give the codebreaker not one letter but two.

Other very notable contacts are: e follows b more than 30 percent of the time. The same for e following h, u follows j, e follows k, h follows t, e follows v, e follows z. Certainly, these combinations should be avoided since the codebreaker is looking for just such pairs of letters.

A reverse chart can be made of high contact with the letter preceeding: d, for example, is preceeded by e more than 30 percent of the time. The same is true for all of the following: o preceeds f, n preceeds g, t preceeds h, m preceeds j, s preceeds q, e preceeds r, o preceeds u, e preceeds v, e preceeds x more than three-fourths of the time, i preceeds z more than half the time.

INDEX